A Volume in The Laboratory Animal Pocket Reference Series

The Laboratory
XENOPUS sp.

The Laboratory Animal Pocket Reference Series

Series Editor
Mark A. Suckow, D.V.M.
Freimann Life Science Center
University of Notre Dame
South Bend, Indiana

Published Titles

The Laboratory Canine

The Laboratory Cat

The Laboratory Guinea Pig

The Laboratory Hamster and Gerbil

The Laboratory Mouse

The Laboratory Nonhuman Primate

The Laboratory Rabbit

The Laboratory Rat

The Laboratory Small Ruminant

The Laboratory Swine

The Laboratory XENOPUS sp.

A Volume in The Laboratory Animal Pocket Reference Series

The Laboratory
XENOPUS sp.

Sherril L. Green

DVM, PhD, Diplomat ACVIM

Professor and Chair, Department of Comparative Medicine
Director, Stanford University Veterinary Service Center
Stanford University School of Medicine
Stanford, California

CRC Press
Taylor & Francis Group
Boca Raton   London   New York

CRC Press is an imprint of the
Taylor & Francis Group, an **informa** business

CRC Press
Taylor & Francis Group
6000 Broken Sound Parkway NW, Suite 300
Boca Raton, FL 33487-2742

© 2010 by Taylor and Francis Group, LLC
CRC Press is an imprint of Taylor & Francis Group, an Informa business

No claim to original U.S. Government works

Printed in the United States of America on acid-free paper
10 9 8 7 6 5 4 3 2 1

International Standard Book Number: 978-1-4200-9109-0 (Paperback)

Library of Congress Cataloging-in-Publication Data

Green, Sherril L.
    The laboratory Xenopus sp. / Sherril L. Green.
      p. cm. -- (The laboratory animal pocket reference series)
    Includes bibliographical references and index.
    ISBN 978-1-4200-9109-0 (alk. paper)
    1. Frogs as laboratory animals. 2. Xenopus. I. Title.

SF407.F74G74 2010
639.3'789--dc22
                                    2009032045

Visit the Taylor & Francis Web site at
http://www.taylorandfrancis.com

and the CRC Press Web site at
http://www.crcpress.com

contents

preface xi
acknowledgments xv
about the author xvii

1 important biological features 1

introduction 1
habitat and geography 2
behavior 4
anatomic and physiologic features 6
    General Features 6
    Integument 7
    Sensory 9
    Gastrointestinal and Excretory 10
    Reproduction 11
    Respiratory 12
    Cardiovascular 14
    Thermoregulation 16
    Longevity 17
    Aestivation 17
references 17

2 husbandry 19

introduction 19
macroenvironment 20
microenvironment 22
    Housing Systems and Water Sources 22
    Static/Closed Systems 24

Flow-Through Systems..25
Modular/Recirculating Systems.......................................27
Filtration Systems and UV Water Sanitation Systems 29
Mechanical Filtration...29
Biological Filtration...31
Chemical Filtration..33
Ultraviolet Sterilization 34
water quality 35
pH 36
Alkalinity..37
Temperature 38
Conductivity 40
Hardness...40
Ammonia (NH₃)..41
Nitrate and Nitrite..42
Chlorine/Chloramines 42
Dissolved Oxygen..43
Total Dissolved Gases—Percent Saturation (TDG%) 44
Carbon Dioxide...44
Water Clarity 45
Miscellaneous Water Toxicants 45
Monitoring Water Quality..45
stocking density 47
photoperiod 48
nutrition...51
Types of Food 51
Frequency of Feeding..53
How Much to Feed 53
sanitation..54
environmental enrichment 55
identification 56
transportation of *Xenopus*...57
record keeping...58
references..59

3 management...63

regulations and regulatory agencies 63
occupational health and safety: injury and zoonotic risks 64
references..66

4 veterinary care 69

physical examination 69
quarantine 70
clinical problems 71
    Bacterial Infections... 74
        Red leg syndrome 74
        Chryseobacterium (formerly called
        Flavobacterium) spp....................................75
        Mycobacterium spp. 80
        Chlamydia spp. 81
    Viral Infections 84
        Ranavirus 84
        Lucke herpesvirus..85
    Fungal Infections...85
        Batrachochytrium dendrobatidis 85
        Saprolegnia spp. and other water molds....87
    Parasitic Infections 90
        Pseudocapillaroides xenopi (*Capillaria xenopodis*) 90
        Rhabdias (strongyloid lungworms)............92
        Cryptosporidia 93
        Acariasis (mites)...93
noninfectious diseases and conditions...................................94
    Dehydration and Desiccation 94
    Gas Bubble Disease 95
    Chlorine/Chloramine Toxicities...........................98
    Neoplasia 100
    Rectal and Cloacal Prolapses 100
    Gout 101
    Skeletal Deformities... 101
    Bite Wounds 102
    Ovarian Hyperstimulation Syndrome.................. 103
    Thermal Shock 104
    Poor Egg Production, Poor Egg Quality 104
diagnosis of infectious diseases in laboratory *Xenopus*....... 106
general comments on the treatment of infectious diseases... 107
treatment of general trauma and abrasions 109
anesthesia... 110
    Tricaine Methanesulfonate (MS-222)............................ 110
    Benzocaine Gel... 111

Ketamine 112
Eugenol (Clove Oil) 112
Propofol 112
aseptic surgery 112
analgesics and post-operative care 114
euthanasia 115
references.. 117

5 experimental methodology.. 125

catching and handling *Xenopus*.. 125
compound administration techniques................................. 126
blood sample collection and interpretation........................ 128
Blood Sample Collection.. 128
Processing the Samples 130
Xenopus hemocytology: characteristics............................. 131
interpretation of the hemogram and serum clinical
chemistries... 134
egg/oocyte collection 134
Surgical Laparotomy for Egg/Oocyte Harvest
from *Xenopus*.. 137
raising *Xenopus* tadpoles 138
necropsy... 140
Necropsy Equipment... 140
Necropsy Technique 141
references.. 144

6 resources... 147

organizations.. 147
electronic resources.. 148
Electronic Resources for *X. tropicalis*........................... 148
Electronic Resources for *X. laevis*................................. 149
Guidelines and Protocols for Harvesting Oocytes 149
Additonal Guidelines... 149
periodicals.. 150
books 151
vendor contact information 151
Microchips.. 151
Carriers Who Will Ship *Xenopus* 152
Frog Suppliers (Frogs and Food)................................... 152
Modular Housing for Laboratory *Xenopus* 153

Sanitation Supplies.. 153

Water Quality Sensors 153

Water Filtration Systems.. 154

Water Test Kits (Spectrophotometric)............................ 154

Water Purifiers.. 154

taxonomy and natural history 154

anatomy and histology 154

physiology 155

ontogeny... 155

genetics 155

medicine and surgery 156

xenopus listservs... 156

index... 157

preface

The South African clawed frog, *Xenopus laevis*, has a long history in laboratory research and classroom teaching. In 1802, French naturalist and zoologist Francois Marie Daudin wrote the first description of what he called *Bufo laevis*, or "Crapaud lisse" (smooth toad). Over the following decades, taxonomists and anatomists continued to study the unusual species, which was neither a typical frog nor a typical toad. In 1890, J.M. Leslie gave the earliest account of the breeding habitats of the species that had become known as *Xenopus laevis*. That name stuck and over the next 30 years, *Xenopus* rose to dominance in research, imported from Africa to laboratories in Europe and North America. Between 1930 and 1960, *Xenopus* became the leading animal model in developmental biology, endocrinology and biochemistry. In the 1940s, the species gained widespread use in hospital laboratories as a means of detecting human pregnancy. (Injection of urine from a woman results in *Xenopus* egg-laying if the hormone associated with human pregnancy, human chorionic gonadotrophin, is present.) In the early 1980s, the Frog Embryo Teratogenesis Assay *Xenopus* (FETAX) was developed and is now the standard laboratory system that tests the toxicity of environmental pollutants and other substances by evaluating their effects on embryonic mortality, growth rate, and malformation. Today, *Xenopus* is a major non-mammalian laboratory animal model in vertebrate embryology, toxicology, cellular biology, physiology, biochemistry, and biomedical research.

Xenopus are especially suitable for the laboratory because they are hardy, long-lived, and readily adaptable, and because gametogenesis can be hormonally induced to occur year round. Thus the oocytes, eggs, and embryos harvested from *Xenopus* can provide scientists with a continuous source of material for research. Eggs and oocytes (fully-grown egg precursor cells) collected from *Xenopus* are plentiful and quite large (~1.0 to 1.2 mm) and are used to prepare cell-free cytoplasmic extracts to study a variety of biochemical and cellular processes—the mitotic cell cycle, molecular signaling pathways, nuclear assembly, nuclear transport, and the transcription, translation, and expression of mRNAs—and for the analysis of ion channels via electrophysiological recording and expression cloning. Eggs and embryos are also readily suitable for microinjection and microsurgery and are used to study cell fate, axis formation, migration, and morphogenesis of the developing vertebrate.

X. *tropicalis*, a smaller cousin of X. *laevis*, shares all of the utility of X. *laevis* as an animal model but offers the additional advantage of being a diploid species that readily lends itself to genomic sequencing. X. *laevis* is allotetraploid: that is, many genes are represented by extra copies that may or may not be functional (this feature makes creating X. *laevis* mutants and analyzing gene regulation difficult). X. *tropicalis* has the additional advantages of having a comparatively shorter generation time (in 3 to 4 months of morphogenesis they will reach sexual maturity, versus 1 to 2 years for X. *laevis*) and a smaller genome (10 pairs of chromosomes compared with 18 for X. *laevis*). As transgenic X. *tropicalis* become more widely used and available, the use of this species in laboratory research will expand.

Because reporting methods vary from country to country, the exact number of *Xenopus* used in research worldwide is unavailable. However, between 1998 and 2009, the number of published studies making use of *Xenopus* as cited on *PubMed* increased five fold. Just two decades ago, research animal facilities were unlikely to house more than a few hundred frogs. Today, maintaining very large populations of *Xenopus* (several hundred to several thousand frogs) in a laboratory environment is increasingly common.

Despite the popularity of *Xenopus*, there is a lack of standardization and a paucity of hypothesis-driven research regarding their optimal care, husbandry, and housing. Much of what we currently know about caring for large numbers of this species in captivity has been learned through trial and error, time-honored laboratory techniques, and contributions from hobbyists, wildlife specialists, and zoos. Most laboratory animal veterinarians have limited formal training or experience with *Xenopus* and must often learn the basics of aquaculture, amphibian medicine, and experimental methodology while on the job.

This handbook is written to assist research scientists, technicians, animal caretakers, and particularly the laboratory animal veterinarians charged with the care of laboratory *Xenopus.* The book is organized into six chapters: Important Biological Features (Chapter 1), Husbandry (Chapter 2), Management (Chapter 3), Veterinary Care (Chapter 4), Experimental Methodology (Chapter 5), and Resources (Chapter 6). This handbook is not intended to be an exhaustive review of *Xenopus* biology or of research methodology, but rather a quick reference for concise and basic information. Individuals performing procedures described in this handbook should receive proper training, keeping in mind the humane use and the welfare of a species that was once considered of a "lower" order. The products and protocols cited throughout this book are not an endorsement over those not listed, but rather are intended as starting reference points.

It is my sincere hope that those who manage laboratory *Xenopus* colonies ranging in size from just a few frogs to several thousand will find this manual useful.

Sherril L. Green, DVM, PhD
Professor, Department of Comparative Medicine
Stanford University School of Medicine

acknowledgments

I would like to thank my colleagues—the many veterinarians, veterinary nurses, animal husbandry staff, and research laboratory personnel—for their contributions to this book. I am grateful for your support. I would also like to thank the talented veterinary pathologists who have contributed so much to our understanding of conditions and diseases affecting laboratory *Xenopus*: Drs. Donna Bouley, Kim Waggie, Corrine Davis, and Richard Luong. Your willingness to "wade right in" is greatly appreciated. Thank you to Barry Lifland, a gifted microbiologist whose expertise is beyond compare.

Lastly, thank you to my husband and my children for your patience. With your love and support, all things are possible.

about the author

Sherril L. Green, DVM, PhD, Diplomat ACVIM, is a professor and the Chair of the Department of Comparative Medicine/Director of the Veterinary Service Center at Stanford University School of Medicine. She earned her veterinary degree from Louisiana State University School of Veterinary Medicine, completed an internship in large animal medicine and surgery at the University of Missouri–Columbia and a residency in large animal medicine at the University of Florida College of Veterinary Medicine. She obtained a doctorate degree in neurobiology from the University of California–Davis before she joined the faculty at Stanford University. Dr. Green has a long-established interest in the husbandry and veterinary care of laboratory *Xenopus.*

important biological features

introduction

- *Xenopus* belong to the order *Anuran* (*anuran* means "tailless") and to the family of tongueless frogs known as *Pipidae*. The family Pipidae comprises two subfamilies: the Pipinae and the Xenopodinae (genera *Silurana* and *Xenopus*).

- More than 20 subspecies of Xenopodinae have been identified to date and it is likely that more remain to be discovered. New *Xenopus* species can evolve through allopolyploidization, a type of genome duplication that can result from hybridization between various species.

- *X. tropicalis* and *X. epitropicalis* now form the genus *Silurana*. All remaining species in the Pipidae family, including *Xenopus laevis*, belong to the genus *Xenopus*.

- *Xenopus laevis* (also known as the South African clawed frog, or the common plantanna) and the smaller, but closely related *Xenopus tropicalis* (also known as the Western clawed frog) are most commonly used in research (**Figure 1**).

- *Xenopus* is Latin for "peculiar or strange foot," referring to this species' large, webbed, five-toed, three-clawed rear foot (**Figure 2**). *Laevis* means "smooth," referring to the species' smooth skin.

- Important differences between *X. laevis* and *X. tropicalis* are summarized in **Table 1**.

Figure 1 Side by side comparison of *Xenopus laevis* and *Xenopus tropicalis*. Note *X. tropicalis* are considerably smaller than *X. laevis*. (From xlaevis.com. With permission.)

- Selected external and internal anatomical features of *Xenopus* are shown in **Figure 3**. Note the males are much smaller than the females (**Figure 4**).

habitat and geography

- *Xenopus* are native to Africa, in a geographic range extending from the southern Cape of South Africa to northeastern Sudan and to the west into Nigeria (Tinsley and McCoid 1996). Extensive populations are also found in Kenya, Cameroon, and the Democratic Republic of Congo.
- As a result of international trade and commerce and their release or escape as pets or from research laboratories, they can now be found around the world. *Xenopus* are an invasive species and are considered a threat to native wildlife in the United States, Great Britain, Portugal, Chile, and Germany.
- *Xenopus* are fully aquatic and require water at all stages of their life cycle. They prefer still, muddy-bottomed pools of

Figure 2 The hind feet of *Xenopus*, showing the claws on the inner three toes.

TABLE 1 IMPORTANT MAJOR DIFFERENCES BETWEEN
X. TROPICALIS AND *X. LAEVIS*

	X. laevis	*X. tropicalis*
Ploidy	Allotetraploid	Diploid
N	18 chromosomes	10 chromosomes
Genome size	3.1×10^9 bp	1.7×10^9 bp
Temperature optima	16°–22°C	25°–30°C
Adult size	10 cm	4–5 cm
Egg size	1–1.3 mm	0.7–0.8 mm
Eggs/spawn	300–1000	1000–3000
Generation time	1–2 years	< 5 months

Source: http://faculty.virginia.edu/xtropicalis/overview/intro.html.

warm fresh water: ponds, lakes, ditches, swamps, and watering holes, either man-made or natural. However, this readily adaptable species has also been found in riverbeds and streams, in salt water, and in areas where ambient temperatures vary from freezing to desert-hot.

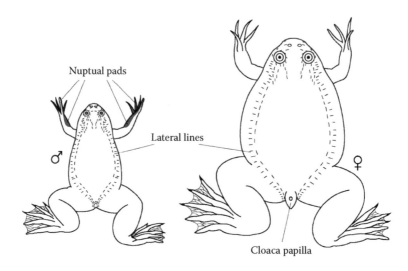

Nuptual pads

Lateral lines

♂

♀

Cloaca papilla

Figure 3 Selected external anatomical features of *Xenopus.*

- They have been known to migrate miles across land in search of water during droughts or when conditions in a particular location become unfavorable (Tinsley and McCoid 1996). They move on land by leaping forward and landing flat on the belly.
- Natural predators include birds, fish, otters, and humans. In Africa, *Xenopus* are farmed for human consumption.

behavior

- In the wild, *Xenopus* live in large groups or colonies. They establish hierarchies within their living groups, and they are territorial.
- *Xenopus* generally remain still in the water, hanging perpendicular at the water's surface or sitting on the bottom of the pond, moving only to catch prey or escape from a threat.
- *Xenopus* are nocturnal and are most active (breeding, feeding) after dark (Elepfandt 1996a).
- They will struggle when captured. Their main defense against predators is escape, using their powerful hind legs to dart forward or backward to seek cover in the mud or among rocks and plants.

Figure 4 Side by side comparison of a female *X. laevis* (the larger frog) and a male. The males of *X. laevis* and *X. tropicalis* are typically 10–30% smaller than the females. The smaller male is pale in comparison because he has been housed in a white, opaque tank. The female was housed in a dark green, pond-style tank.

- *Xenopus* males lack a true vocal sac, but they vocalize and produce a mating call by rapid contractions of the intrinsic laryngeal muscles (Elepfandt 1996b). The male mating call sounds like alternating long and short trills. The female responds with either an acceptance call (a rapping sound) or a rejection call (a slow ticking sound) (Tobias et al. 1998). These sounds can best be heard underwater with the aid of a hydrophone. To hear underwater recordings of the male *Xenopus laevis* advertisement call, go to http://www.californiaherps. com/frogs/pages/x.laevis.sounds.html.

- *Xenopus* are carnivorous predators with an indiscriminate, voracious appetite (Tinsley et al. 1996). They are also scavengers and will readily eat carrion, but they prefer live food: fish, birds, slugs, worms, and other aquatic invertebrates. Stomach content examinations of wild *Xenopus* will also include plant matter, insects, and plankton.

- They are cannibalistic, as are many Anuran species: larger frogs will eat froglets, tadpoles, and their own eggs.
- *Xenopus* feed in a frenzy, using their front feet to shovel food into the mouth and their hind feet to shred.

anatomic and physiologic features

General Features

- *Xenopus* is a smooth, dark-skinned frog with a flattened body and a small head with a blunt, rounded snout.
- *Xenopus* have short forelimbs and unwebbed forefeet with four digits, which are used to direct food into the mouth.
- The hind limbs are larger and more muscular, and are used to propel the frog through the water.
- The hind feet are fully webbed, and the inner three of the five toes on the hind limbs have sharp, black claws (**Figure 2**). These are used to dig insects from the mud, shred food items, and to dig in the mud during aestivation.
- In this species, the proper term for the body cavity (where the gut and other organs are located) is the **coelomic cavity**. The coelomic cavity forms from the mesoderm during embryonic development. Fluid and egg mass accumulation within the distendable coelomic cavity can result in great increases in body size.
- *Xenopus* are technically ectotherms, meaning the body temperature is dependent on the environment. They cannot increase their metabolism to increase their body temperature.
- Sexually mature adult female *X. laevis* range from ~9 to 14 cm snout-to-vent length (SVL) at 2 to 3 years of age, weighing from 20 to 150 grams, although they can grow much larger (**Figure 4**). Sexually mature males are typically 10–30% smaller than the females (Kobel et al. 1992).
- Sexually mature female *X. tropicalis* range in size from 4 to 5 cm SVL and 10 to 50 grams, depending on their age and genetic background. *Xenopus tropicalis* females are also much larger than the males.

Integument

- *Xenopus* skin is permeable and slimy. The protective slime, or mucous layer, serves as a barrier against pathogens and abrasions. Serous glands on the head and shoulder synthesize and secrete peptides, including magainins, caeruleins, and other substances that have antibacterial, anti-viral, and antifungal properties (Kreil 1996). Care should be taken when working with this species not to damage or remove the slime layer.

- The slime layer and the dead skin of *Xenopus* are regularly shed, eaten, and replaced. Shedding begins on the dorsum at the head and between the shoulders. *Xenopus* can be seen scraping off shed skin with their hind feet.

- The skin of *Xenopus* loosely adheres to the underlying structures by thin, transparent septae that divide the large subcutaneous space into discrete sacs (Carter 1979). These spaces are called **lymphatic sacs**. (**Figure 5**) Medications injected

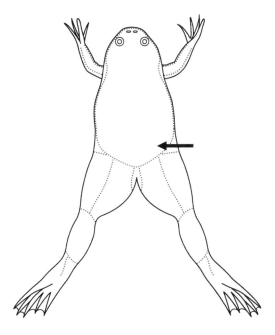

Figure 5 A dorsal view of an adult *Xenopus* showing the subcutaneous sacs and the septa (dotted lines). The arrow points to the dorsal lymphatic sac, the most common location for administration of subcutaneous medications.

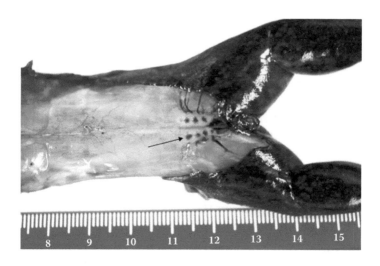

Figure 6 The lymphatic hearts. (Photo courtesy of Dr. Donna Bouley.)

subcutaneously into the lymphatic sacs drain with the lymph into the venous system, aided by small contractile structures called **lymphatic hearts** (**Figure 6**). In a freshly dissected specimen, the hearts can often be seen actively beating.

- The **lymphatic hearts** can be found just beneath the skin on the dorsum. The anterior hearts are located on each side just beneath the scapula (these may be absent in mature adult frogs), and the posterior hearts (which vary in number from one to six pairs) are located on each side of the urostyle (a long unsegmented bone that represents a number of fused vertebrae and forms the posterior part of the vertebral column of frogs and toads) (**Figure 6**). The anterior and posterior lymphatic hearts drain fluid from the subcutaneous sacs into the bloodstream via the anterior vertebral vein and the renal portal veins, respectively.

- The subcutaneous space and the posterior lymphatic sacs, in conjunction with the posterior lymphatic hearts, the renal portal vein, and the kidney, form a mechanism for rapidly excreting body water. Water influx via the skin is 2% of the body mass per hour in adult *Xenopus* (Boutilier et al. 1992). The subcutaneous space can greatly distend and fill with fluid as a water reserve as needed, or as result of a pathologic process (renal failure, sepsis, or heart failure).

- *Xenopus* skin is highly permeable to gases, solutes, and many other water constituents, including toxicants. Thus *Xenopus* depend on a nonpolluted source of water.

- *Xenopus* exchange water and electrolytes through the skin to protect against dehydration and to maintain normal plasma osmolality. Oral uptake and drinking play an unimportant role in *Xenopus* hydration and osmoregulation.

- *Xenopus* dorsal skin color can be pale olive green to brown, with mottled markings. Chromatophores, the cells containing pigment, are under hormonal control and allow *Xenopus* to change color to better adapt to their habitat. *Xenopus* housed in white or opaque tanks are often pale green in color, while those housed in dark green or black tanks will be dark green, almost black. The degree of mottling or stippling can also depend on the genetic background. The underbellies of healthy *Xenopus* are usually yellowish-white.

- Because melanin pigments in lower vertebrates are often found in locations other than the skin, albino *Xenopus* are often used to study the cellular and biochemical structure and functions of this system. Depending on the age and the strain, albino *Xenopus* can vary in color from white to pale yellow. Some albino strains are bred by hobbyists for their reticulated skin patterns.

Sensory

- *Xenopus* have small eyes on the top of the head, which are well placed for overhead vision (to detect food and potential predators at the water's surface) (Elepfandt 1996a). In their natural habitat, *Xenopus* are found in dark, murky waters. Vision underwater is thus not thought to be a vital sensory input.

- The ears are not visible externally, but *Xenopus* have a well-developed auditory system. The tympanic membrane is equipped for detecting high-frequency sounds, for underwater communication between individuals, and for detecting low frequency sounds through the bone structures of the skull.

- *Xenopus* also have a keen sense of smell. Indeed, the bulk of the cerebral cortex consists of olfactory lobes. The nostrils are on top of the head at the end of the snout, positioned so that animals can smell food (blood and rotting carrion, especially) in the water *and* in the air. *Xenopus* lack nasolacrimal ducts.

- Rows of hair cell receptors on the head and body of *Xenopus* form the lateral line system. The lateral line system runs along the sides of the body (the dorsum and the ventrum) and around the head and eyes as a symmetrical pattern of small vertical "stitches," as shown in **Figure 3**.

- One portion of the lateral line system is a mechanoreceptive system that responds to water movements and vibrations. The other portion of the lateral line system consists of electro-receptors sensitive to the presence of electrical fields in the water (Wilczynski 1992).

- The lateral line system is innervated by long branches of special cranial nerves that transmit information to the nuclei in the medulla. The lateral line system is used to detect prey and predators via changes in water movement (Elepfandt 1996a).

Gastrointestinal and Excretory

- *Xenopus* lack a fleshy, moveable tongue, and they lack teeth, although small spike-like serrations can be seen and felt along the inside of the lower jaw.

- *Xenopus* swallow their prey whole or in shredded bits. *Xenopus* are not thought to drink significant volumes of water. They maintain hydration primarily by the exchange of water and ions through their skin.

- *Xenopus* can regurgitate the stomach and use their feet to wipe off the mucosa to remove indigestible or toxic substances (Tinsley et al. 1996). They may regurgitate food or regurgitate their stomach and expel the contents if they are stressed or disturbed too soon after eating.

- The oral cavity, esophagus, stomach, and intestine are separated by sphincters, but in *Xenopus* further distinct sections of the intestine are not grossly obvious.

- A gall bladder and pancreas are present, and the large liver is bilobate. Melanin is a pigment normally found in the frog liver and can sometimes give the liver a mottled appearance. The liver plays little role in processing nitrogen for excretion since ammonia, the primary nitrogen waste form in *Xenopus*, is freely diffused into the environment through the skin and via excretion by the kidneys.

- Large, yellow-gold or pale yellow finger-like fat pads are present in the coelomic cavity of *Xenopus* and are a source of energy during high metabolic activity (egg laying and gametogenesis) or during times of diminished food supply (Tinsley and McCoid 1996).

- Amphibians, including *Xenopus*, have a more primitive, mesonephric rather than a metanephric kidney as seen in mammals. *Xenopus* are primarily ammonotelic, that is, they excrete nitrogen waste in the form of ammonia. When environmental conditions are adverse (drought, for example), *Xenopus* become ureotelic (they excrete waste as urea).

- *Xenopus* cannot concentrate urine beyond the normal plasma osmolarity, but urine is collected in a bladder, as in other species. The urinary bladder in *Xenopus* is very small.

- *Xenopus*, like birds and reptiles, possess a common waste-collecting chamber called a **cloaca**. Products from the urinary system, the gastrointestinal system, and the reproductive tract all empty into the cloaca before exiting the cloacal opening.

Reproduction

- Breeding in the wild is seasonal, with peak production periods in temperate climates starting in early spring and lasting through early fall. In ideal conditions, gravid females may lay multiple clutches during a single season (Tinsley et al. 1996).

- Sexually mature females have a large, readily visible cloacal flap and may produce 500 to 3,000 plus eggs per clutch. In a laboratory environment, most female *X. laevis* lay (or can be hormonally induced to lay) several hundred to a thousand or more eggs per collection cycle. Captive *X. tropicalis* can lay 1,000 to 3,000 eggs.

- Sexually mature male *Xenopus* have darkened nuptial pads on the inner forelegs and fingers (see **Figure 3**). Nuptial pads are most visible during the mating season.

- During amplexus, a 3- to 4-hour mating embrace, the male grasps the female from behind (**Figure 7**). Fertilization occurs externally, with the male releasing sperm over the egg mass.

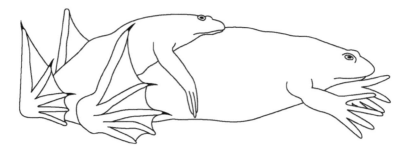

Figure 7 *Xenopus* in amplexus, a mating embrace where the smaller male grasps the female from behind.

- Sticky, jelly-covered *Xenopus* eggs are bi-colored (**Figure 8**) and are characterized by a dark "animal pole" and pale yellow to white "vegetal pole" (the yolk).
- Once fertilized, the eggs rotate so that the animal pole is the upward-facing pole. The eggs grow into gill-breathing tadpoles, which filter feed (Wassersug 1996).
- Tadpoles are translucent to pale yellow to white, with whiskers at the corners of the mouth and a tail that ends in a filament (**Figure 9**). *Xenopus* tadpoles swim at an angle, with head down, using the tail to move food downward.
- The tadpole metamorphoses into a lung-breathing froglet (a small, tailless frog with limb buds). The entire developmental process (starting from a fertilized egg developing into a small juvenile froglet) takes about 6 to 8 weeks for *X. laevis* and 3 to 6 weeks for *Xenopus tropicalis*. The development cycle of *Xenopus* is shown in **Figure 10 A and B**.
- After metamorphing into a froglet, tadpoles lose their gills and convert to a primarily carnivorous diet.
- Images of stages of *Xenopus* development, based on the Nieuwkoop and Faber (1994) Normal Table of *Xenopus laevis* (Daudin), can be seen at http://www.xenbase.org/anatomy/static/NF/NF-all.jsp and http://www.engr.pitt.edu/ldavidson/NieuwkoopFaber/Frame1.html.

Respiratory

- Adult *Xenopus* are lung breathers. They rely on well-developed lungs and the ability to rise to the water's surface to breathe to exchange gases.

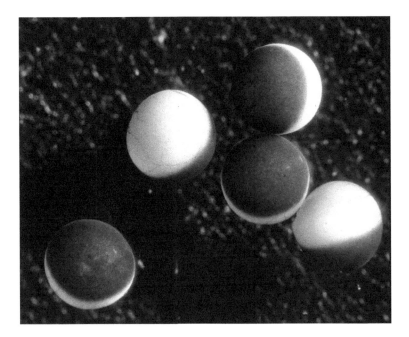

Figure 8 Bi-colored *Xenopus* eggs are characterized by a dark "animal pole" and pale yellow to white "vegetal pole" (the yolk). Fertilized eggs have rotated so that the animal pole (the dark half of the egg) is facing upward. The diameter of the eggs is ~1.2 mm. (Photo courtesy of Dr. Erwin Sigel.)

- *Xenopus* have variable, intermittent breathing patterns characterized by breath-holding while diving followed by rising to the surface for short breaths of air (Boutilier 1984). Alternately, *Xenopus* may rise and linger at the water's surface for prolonged bouts of breathing.

- The trachea is very short and bifurcates into simple lungs that lack alveoli and are not multilobed.

- *Xenopus* do not have ribs, and they lack a diaphragm. During inhalation, air is drawn into the buccopharyngeal cavity and forced into the lungs by contraction of the floor of the buccal cavity (the glottis opens, the nares are shut). *Xenopus* expulse lung gas by contracting the flank musculature. Exhalation is also aided by elastic recoil and pulmonary smooth-muscle effects.

- In addition to the gas exchange that occurs in their lungs, *Xenopus* also exchange some gases through the skin, primarily

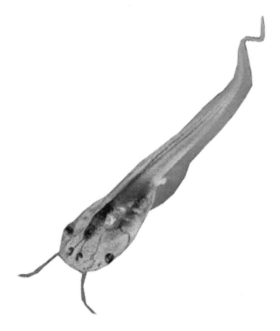

Figure 9 *Xenopus* tadpoles are translucent, with whiskers at the corners of the mouth and a tail that ends in a long filament. This tadpole represents ~stage 51 based on the Nieuwkoop and Faber (1994) Normal Table of *Xenopus laevis* (Daudin).

carbon dioxide (CO_2). Intake and exchange of oxygen via the skin alone is not sufficient to sustain the animal.

Cardiovascular

- Larval *Xenopus* have a two-chambered heart (the atrium and the ventricle), while adult *Xenopus* have a three-chambered heart (a right and a left atrium and a single ventricle).
- The left atrium is smaller than the right, and the interatrial septum between the two is incomplete.
- The ventricular trabeculae are thick, and the superficial vasculature is darkly pigmented and hard to see.
- The heart of *Xenopus* has been extensively studied. As an organ, either *in vivo* or *in vitro*, it will continue beating for many hours—*in vivo* even after the frog has been euthanized, has lost withdrawal and righting reflexes, and has stopped breathing.

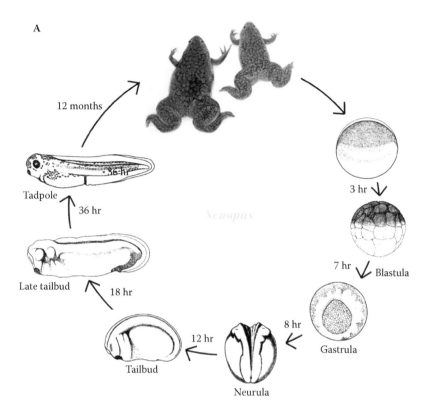

A

12 months

Tadpole

36 hr

Late tailbud

18 hr

Tailbud

12 hr

Neurula

8 hr

Gastrula

7 hr

Blastula

3 hr

Figure 10 A) The developmental cycle of *Xenopus* (Image from *Xenbase* at http://www.xenbase.org/common/. With permission.) B) *X. laevis* tadpoles at ~stage 51 (tadpole on the upper right), stage 56 (middle tadpole), and stage 61 (the froglet), based on Nieuwkoop and Faber (1994) Normal Table of *Xenopus laevis* (Daudin). Note the hind limb buds develop first. The tail will be resorbed by ~stage 66, a process that takes about 4 to 5 days to complete. (Photo courtesy of Dr. Bob Denver.)

- Amphibian physiology is temperature dependent. The heart rate for adult *Xenopus* is ~8 beats/min at 2°C and is ~40–60 beats/min at 25°C (Taylor and Ihmied 1995).

- The blood plasma osmolality of a healthy adult *Xenopus* living in fresh water ranges from 200 to 233 mOsm/kg, depending on the water pH and water temperature. This is lower than the plasma osmolality of mammals (250 to 300 mOsm/kg) (Boutilier et al. 1992).

B

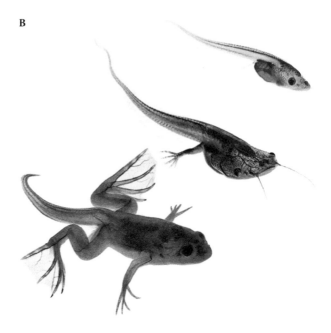

Figure 10 (continued).

Thermoregulation

- Amphibians are ectotherms and so cannot raise their body temperature by producing metabolic heat. Their main source of heat for body temperature regulation is the environment. They rely to a lesser degree on behavioral thermoregulation (selection of water temperatures, seasonal cycles of breeding activity, etc.), internal adjustments of the cardiovascular system, thyroid-induced changes in metabolism, muscular activity, alterations of central and peripheral neuronal responses, and their ability for tolerance and acclimation to environmental factors (temperatures, photoperiod, water availability, etc.) (Hutchison and Keith Dupré 1992).

- Because *Xenopus* invest so little energy in heat production, their overall energy requirements can be incredibly low. Thus they can tolerate prolonged starvation and persist in food-poor habitats.

- *Xenopus* are adaptable to a wide range of water temperatures. However, they are susceptible to heat and cold shock. Rapid introduction into cold or hot water (water just 2° to 5°C colder or warmer than the primary enclosure water) can result in thermal shock and death.

Longevity

- *X. laevis* have been known to survive 15 or more years in the wild (Tinsley et al. 1996) and 25 to 30 years in captivity.
- In a research animal facility, 3 to 6 years is typical of female frogs kept solely for egg production. After 6 years of age, egg production in captive females tends to decline, although several research institutions have kept captive good egg-producing females for 12 years or longer.

Aestivation

- In the wild, *Xenopus* will aestivate (enter a state of torpor or dormancy during unfavorable conditions: cold or drought) if the water temperature drops below 8°C or if their water sources dry out (Tinsley et al. 1996). During aestivation, *Xenopus* will burrow into the mud, breathing by way of a slime-coated tube-like tunnel to the surface, and remain dormant for up to a year or more or until environmental conditions become suitable again.

references

Boutilier, R. G. 1984. Characterization of the intermittent breathing pattern in *Xenopus laevis. J. Exp. Biol.* 110:291–309.

Boutilier, R. G., D. F. Stiffler, and D. P. Toews. 1992. Exchange of respiratory gases, ions, and water in amphibious and aquatic amphibians. In *Environmental physiology of the amphibians*, ed. M. F. Feder, and W. W. Burggren, 81–124. Chicago: University of Chicago Press.

Carter, D. B. 1979. Structure and function of the subcutaneous lymph sacs in the Anura (Amphibia). *J. Herpetol.* 13:321–27.

Elepfandt, A. 1996a. Sensory perception and the lateral line system in the clawed frog, *Xenopus*. In *The biology of Xenopus*, ed. H. R. Tinsley and H. R. Kobel, 97–120. New York: Oxford University Press.

Elepfandt, A. 1996b. Underwater acoustics and hearing in the clawed frog, *Xenopus*. In *The biology of Xenopus*, ed. H. R. Tinsley and H. R. Kobel, 177–94. New York: Oxford University Press.

Hutchison, V. H. and R. Keith Dupré. 1992. Thermoregulation. In *Environmental physiology of the amphibians*, ed. M. E. Feder and W. W. Burggren, 206–49. Chicago: University of Chicago Press.

Kobel, H. R., C. Loumont, and R. C. Tinsley. 1992. The extant species. In *The biology of Xenopus*, ed. R. C. Tinsley and H. R. Kobel, 9–34. New York: Oxford University Press.

Kreil, G. 1996. Skin secretions of *Xenopus laevis*. In *The biology of Xenopus*, ed. H. R. Tinsley and H. R. Kobel, 263–278. New York: Oxford University Press.

Taylor, E. W. and Y. M. Ihmied. 1995. Vagal and adrenergic tone on the heart of *Xenopus laevis* at different temperatures. *J. Therm. Biol.* 20(1–2):55–59.

Tinsley, R. C., C. Loumont, and H. R. Kobel. 1996. Geographical distribution and ecology. In *The biology of Xenopus*, ed. R. C. Tinsley and H. R. Kobel, 35–60. New York: Oxford University Press.

Tinsley, R. C. and M. J. McCoid. 1996. Feral populations of *Xenopus* outside of Africa. In *The biology of Xenopus*, ed. R. C. Tinsley and H. R. Kobel, 81–94. New York: Oxford University Press.

Tobias, M. L., S. S. Viswanathan, and D. B. Kelley. 1998. Rapping, a female receptive call, initiates male-female duets in the South African clawed frog. *Proc. Natl. Acad. Sci. U. S. A.* 95:1870–75.

Wassersug, R. 1996. The biology of *Xenopus* tadpoles. In *The biology of Xenopus*, ed. H. R. Tinsley and H. R. Kobel, 195–212. New York: Oxford University Press.

Wilczynski, W. 1992. The nervous system. In *Environmental physiology of the amphibians*, ed. M. E. Feder and W. W. Burggren, 7–39. Chicago: University of Chicago.

2

husbandry

introduction

Protocols describing long-term housing of laboratory *Xenopus* are generally based on the biology of wild amphibians, institutional experience, time-honored laboratory techniques, and recommendations from amphibian hobbyists, zoos, and aquariums. Unfortunately, there is a lack of scientifically tested, evidence-based information on the housing and husbandry requirements necessary to optimize the health and fecundity of laboratory *Xenopus*. *The National Academy of Sciences* offers guidelines regarding the husbandry and care of laboratory *Xenopus laevis*, but the document specifically states the recommendations should be considered tentative, as few tested guidelines are available (National Academy of Sciences 1974). The Council of Europe has also proposed standards for aquatic amphibian housing and husbandry based on the recommendations of experts on amphibians (Council of Europe 2003). Hilken et al. (1995) tested the growth rates of *Xenopus laevis* under different laboratory rearing conditions, but the report is limited by the small number of animals used in each testing group, the short study period, the absence of water quality testing, and the lack of details regarding the frogs' genetic background, age, and sex, and the source of the animals used in the study. Many research groups (Nieuwkoop and Faber 1967; Alexander and Bellerby 1938; Hilken et al. 1995; Sive et al. 2000b; Smith et al. 1991; Wu and Gerhart 1991) have developed successful protocols for the care and rearing of laboratory *Xenopus*, but practices can be quite variable between laboratories. Indeed, a literature review and survey of 66 research laboratories in the United States maintaining *Xenopus laevis* colonies revealed no consistent husbandry or housing protocols (Major and Wassersug 1998).

Xenopus are a fully aquatic species and require water throughout all of their life stages. Water quality and housing are therefore critical to this species. Since the Major and Wassersug report over a decade ago, many animal facilities housing large colonies of *Xenopus* have moved away from using custom-designed, home-built "pond"-style frog tanks and from the labor-intensive, polycarbonate "shoebox" cages originally intended for rodents. Professionally designed and manufactured aquatic housing systems known as *aquaria racks* or *modules* are now commercially available and built specifically for *Xenopus*. Modular systems are expensive, but offer the advantage of a more sophisticated means of water quality management and in some cases a more efficient space-economical method of housing large numbers of *Xenopus*.

Elements of the **macroenvironment** (the space involving the entire room) and the **microenvironment** (the space immediately surrounding the animal, including the housing tanks) critical to the care of laboratory *Xenopus* are presented below.

macroenvironment

- Aquatic amphibian rooms must have sufficient *airflow* to allow surfaces to dry properly and to permit adequate oxygen delivery and carbon dioxide removal. Ten to 15 fresh air changes are recommended (National Research Council 1996).

- Ideally, other species should not be housed in the aquatics rooms with *Xenopus*.

- *X. laevis* and *X. tropicalis* should be housed in different rooms because they differ in water temperature preferences, and their susceptibility to disease varies (particularly to the chytrid fungus: *X. tropicalis* are more susceptible, while *X. laevis* can be silent carriers of the pathogen). Equipment should not be shared between these two species.

- *Xenopus* quarantine rooms should also be separate rooms, with dedicated equipment and services by laboratory personnel who do not go into other aquatic housing rooms.

- The humidity in *Xenopus* housing rooms can be expected to be high, but should be comfortable enough for personnel to work in and low enough to keep structural and equipment damage due to rust, mold, and algae growth to a minimum.

- *Electrical systems* should have ground fault circuit interrupter outlets, and all outlets should be water resistant or fitted with waterproof covers. Cords, power strips, and surge protectors should be placed outside of the splash zones, securely mounted in a dry area, and never suspended directly over an open tank of water.

- *Noise* from construction, machinery and heavy equipment, particularly noise that causes vibrations in the water tanks, will disturb the frogs and should be avoided.

- *Light cycles* that are phased off and on in 12-hour cycles to mimic the frogs' natural habitat are sufficient.

- Water temperature of the frogs' housing tanks will equilibrate to the *room temperature*. Therefore, room temperature should be monitored closely and not deviate more than 6°C in a 24-hour period.

- It is particularly important that floors, walls, and doors in an aquatics facility be easily sanitized, rust-free, and made with water-impervious material (no wood).

- Floors should be made of skid-free material or covered with rubber mats. There should be no standing water on the floor. Floors should gently slope toward the floor drains. The floor drains in aquatics rooms should be oversized to accommodate transient large water flows and covered with mesh to prevent frogs from escaping.

- *Food storage* containers in aquatics rooms should be stored off the floor in a dry and cool (< 20°C) area and kept free from vermin. The food containers should be labeled with the receipt and expiration dates. Live prey should be contained accordingly and should not be stored, but used continuously and the supply replenished.

- Water conditioning chemicals, salts, and cleaning supplies should be stored in waterproof containers away from the feed and the frogs, preferably in a separate room or in a secure cabinet, and labeled accordingly.

TABLE 2 AN EXAMPLE OF *XENOPUS* HOUSING/HUSBANDRY MONITORING LOG

Action	Frequency	1-Jan	2-Jan	3-Jan	...-Jan
Feed frogs	3XW				
Census	D				
Identify and quarantine sick frogs; notify Vet	D				
Submit dead frogs for necropsy; notify Vet	D				
Record water temperature and pH	D				
Record canister filter pressures	D				
Check UV light bulbs	D				
Change UV light bulbs	6–9 M				
Hose off/replace canister filter	W				
Hose off/replace sump tank filter	D				
Clean/vacuum baffles	AN				
Clean/vacuum tanks	AN				
Perform 10% water change	AN				
Check water chemistries	W				
Check dosing pump	D				
Add salt solution to doser pump	AN				
Clean floors	W				
Clean nets, buckets	W				

Note: D, Daily; AN, As Needed; 6–9 M, Every 6–9 Months; W, Weekly; X, Times.

- *Waste materials*, including euthanized frogs, must be disposed of according to local, state, and federal regulations.
- An example of a *Xenopus* housing room's written husbandry/housing monitoring log and equipment maintenance checklist is shown in **Table 2**.

microenvironment

Housing Systems and Water Sources

The three main types of aquatic holding systems—static systems, flow-through systems, and modular/recirculating systems—have been recently reviewed (Koerber and Kalishman 2009). Water used for all types of *Xenopus* housing systems generally comes from one of the following sources: well water, potable tap water (dechloraminated/dechlorinated), or water purified by reverse osmosis and reconstituted (salts added back).

Well water—Well water is used for a source of *Xenopus* housing in some institutions. Fresh well water is usually a safe and reliable source of water for aquatic amphibians; however, it can be high in salt concentration, contaminated with pesticides or herbicides from land runoff, or contain heavy metals, particularly iron and copper. All of these can be harmful to frogs. Supersaturation of total dissolved gases can also occur in well water, especially when the colder well water is rapidly heated or when carbon dioxide from decomposing organic matter or nitrogen gas from denitrification becomes entrained.

Potable tap water—Potable tap water supplied by a municipality will contain disinfectants such as chlorine or chloramines. These chemicals are added at concentrations that are safe for humans but that can stress or kill frogs. Chlorine and chloramines must be removed.

- Chlorine can be removed by passing through a filter containing activated carbon, by "aging" (allowing the water to stand for long periods) such that chlorine evaporates from the surface, or by aeration. Chloramine is much more stable. It must be removed by filtration through a specialized catalytic activated carbon filter and generally requires longer contact times with the carbon. It should be noted that significant amounts of harmful ammonia are produced and released into the water when chloramines are removed by catalytic activated carbon. A healthy biological filter (the microorganisms that convert ammonia to less toxic forms) is imperative so the oxidation of ammonia to the non-toxic forms occurs when chloramine is removed via activated carbon filters.

- In addition, water municipalities often "shock load" water lines with higher concentrations of chlorine/chloramines during warm months or after storms (the intent being to lower the water coliform counts that may pose a public health risk). Higher concentrations of these chemicals may overwhelm carbon filters and result in higher than normal levels of chlorine/chloramines in the water. Contacting the local municipality authorities ahead of time to ask for early notification of any such changes in water treatments is strongly recommended. Frog tank watering systems can be taken off during that time.

- Both chlorine and chloramines in potable tap water can be neutralized by the addition of deaminating chemicals such as Chloram-X™ (AquaScience Research Group, Inc., North Kansas City, MO) or Amquel™ (Kordon LLC, Division of Novalet TIC, Hayward, CA), or many other similar products.

- Heavy metals in potable water must be removed by additional water treatment processes.

- Like well-water, potable water is also susceptible to super-saturation with total dissolved gases, particularly when cold water is heated under pressure or when there is air leaking into the piping system. Degassing systems built into the water delivery and treatment systems may be required.

Purified water—Distilled, reverse-osmosis-treated or deionized water is a water source that is least likely to contain substances harmful to frogs. However, purified water lacks the salts required to safely house the frogs, and frogs should not be placed in puri-fied water that has not been treated with salts. Purified water must be "reconstituted" to appropriate salinity for a fresh water aquatic species such as *Xenopus*. To approximate fresh water salinity (fresh water salinity is usually <0.5 ppt, or 0.5 grams salt/1000 grams water), purified water must be treated with the salts and minerals provided in products such as Instant Ocean® (Aquarium Systems, Inc., Mentor, OH), Bio-Sea Marinemix® (Aqua Craft®, Inc., Hayward, CA), or Coralife® salts (Oceanic Systems, Inc. Oceanic is a registered trademark of Central Garden & Pet, Walnut Creek, CA), [see below], according to the manufacturer's recommendations before the water can be used for the frogs.

Static/Closed Systems

- Fish tank aquaria or polycarbonate rodent shoebox-style cages filled with 4 to 6 liters of potable, dechlorinated/dechloraminated, salt-conditioned water that must be com-pletely changed two to three times a week or more (depend-ing on stocking density and the frequency of feeding) are good examples of static/closed methods of housing *Xenopus* (**Figure 11**).

- Static systems are often used for short-term housing, for quar-antine, or in facilities that maintain relatively few *Xenopus*.

- Static systems are labor intensive because they require fre-quent water changes to maintain good water quality, but they are less costly to set up.

Figure 11 A static housing system, a polycarbonate "shoe-box"-style rat cage, used for laboratory *Xenopus*.

- Complete water changes should be performed at the same time each day, preferably 3 to 4 hours after feeding, when water quality is at its poorest.
- *Xenopus laevis* are powerful jumpers, so shallow, closed systems require secure lids to prevent escape. Secured lids should have air holes and not come into contact with the surface of the water.

Flow-Through Systems

- Flow-through systems allow the water to enter the *Xenopus* holding tank and exit directly to the drain, carrying frog waste and uneaten food with it. Water entry and exit can be continuous (usually at a low volume and a slow flow-through rate), or equivalent to complete or partial water changes (for example, the water is changed 50% over a 20- to 30-minute period every other day). Automatic water exchange systems

Figure 12 Flow-through "pond"-style housing system for laboratory *Xenopus.*

preprogrammed and maintained on a timer are often used in custom-built "pond"-style flow-through systems (**Figure 12**).

- In flow-through systems, the drainage of dirty water and replacement by clean water much improves the overall water quality. Flow-through systems are less expensive to build and maintain than modular housing, but water quality can fluctuate dramatically.

- One disadvantage of flow-through systems is that they require a plentiful water source and thus may not be suitable for water-use restricted facilities. In addition, the flow-through systems' plumbing and float valves can require frequent maintenance.

- Flow-through systems are most often supplied by potable tap-water sources. The water must therefore be filtered to remove particulates, dissolved gases and chlorine/chloramines. In-line sediment and charcoal filters are used for this purpose and require consistent monitoring and replacement.

- In a flow-through system, removal of all of the sediment (feces and uneaten food) from the tank water is often incomplete. Depending on the design of the tank (sloping tank bottoms and good drainage are essential) feces and uneaten food may settle at the bottom of the tank and accumulate. Pool-type vacuums are useful for removal of the accumulated debris.

- Large (300–400 gallon) pond-style flow-through tanks tend to take up a lot of floor space and may not be optimal in animal facilities where square footage for animal housing is constrained.

Modular/Recirculating Systems

- Recirculating *Xenopus* racks or modular systems (**Figure 13**) offer high-efficiency housing and optimal water quality, but they are costly to set up and require a good working knowledge of their maintenance and of water quality management.

- In recirculating modular systems, *Xenopus* can be housed in higher densities within optimal water parameters, as the waste and debris from the water is removed after it has circulated through a series of filters and *before* the water returns to the tanks. The macro environment of the recirculating system can thus be tightly controlled.

- Tank sizes for modular systems range from 15 to 50 gallons, or can be customized, depending on the manufacturer, and built to specifications.

- Modular housing racks usually use purified, reconstituted water as the water source, have their own water-making system and water storage unit, and have a salt doser pump built into the system.

- Modular systems are typically installed with alarm equipment associated with the water quality monitoring system. Preset ranges for the various water quality parameters can be programmed into water monitor systems. When deviations occur, an audible, visual, and pager notification alarm is triggered.

- Overall water usage is conserved with the use of a modular recirculating system. Modular systems are stackable, and in

Figure 13 A stand-alone recirculating/modular housing system for laboratory *Xenopus*.

some cases design and layout of the racks can take up less floor space than the "pond"-style equipment. However, additional room must be made for the water maker and water storage unit, heater, chillers, filters, pumps, and UV light sterilization system.

- The water in modules is usually recirculated and shared among many separate tanks on a rack. Thus a large number of animals are subject to exposure if a problem occurs (disease outbreaks or water quality problems, for example). Sick frogs must be removed immediately from the system.

- Modular systems are not generally built with degasser systems intended to improve water aeration and minimize the likelihood of morbidity and mortality related to gas bubble disease. Entrained air in the water supply of modular systems can be a problem if there is a mechanical problem unless degassers (column aerators, bubblers) are incorporated into the construction (for details, see the section below on gas bubble disease). However, the water flow rate in the modular housing system is very slow (typically, 10% of the module's total water volume is filtered and recirculated per day), thus incoming water is usually free of entrained air unless there is a mechanical problem with the components. Most modular housing suppliers can provide a total dissolved gas monitoring system.

Filtration Systems and UV Water Sanitation Systems

- Good water quality is inherently dependent on aquaria filters and water cleanliness. There are three types of water filters that help maintain water cleanliness: 1) **mechanical filters** such as pads, glass beads, and stones, which provide a way to trap residual debris for collection and removal from the system and provide a substrate for the biological filter to grow (Whitaker 2001); 2) **biological filters**, which are composed of the microorganisms necessary to convert harmful waste metabolites into nontoxic metabolites; and 3) **chemical filters**, such as activated charcoal, which remove dissolved wastes.

- **Ultraviolet lights** in a modular system sanitize the water by sterilization and by killing the burden of harmful microorganisms.

- A basic understanding of the mechanical, chemical, and biological filters and of the ultraviolet sanitation system is critical. Modular-style housing offers the most sophisticated filtration systems for *Xenopus*; the various filtration components in a modular system are shown in **Figure 14**. All components of a modular system are shown, including the water maker, the inline heater/chiller components, the UV sanitation system, and the water quality monitoring equipment.

Mechanical Filtration

- Mechanical filtration removes particulates from the water, including uneaten feed, animal waste, and organic material (bacteria, algae) (Whitaker 2001).

- Mechanical filters in the form of screens, porous media filters (sponges, paper cartridges, beads, and polystrand), or

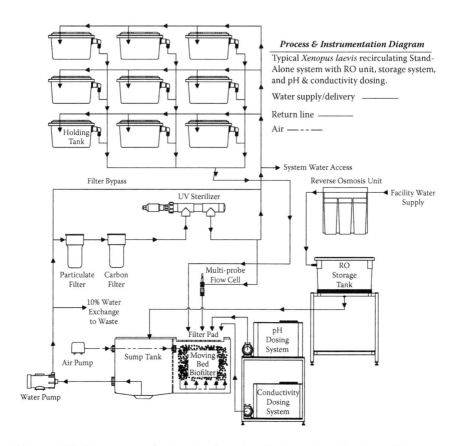

Figure 14 Diagram of a typical recirculating stand-alone *Xenopus* housing system. Reverse-osmosis-treated water flows from the storage tank to a sump tank, en route to the frog holding tanks. The water is adjusted by a pH salt doser, pumped through particulate and carbon filters (during cleaning and replacement of the filters, the filters can be bypassed), and then passes through a UV sterilizer before entering the frog holding tanks. Dirty water is returned from the frog tanks to the sump tank, passing over filter pads and a moving bed biofilter to complete the circuit. The multiprobe flow cell component of the water quality monitoring system sits in the sump tank. Opening an in-line valve to drain 10% of the water allows for replacement with fresh water and is a means of eliminating accumulated nitrate from the system.

granular media filters all provide pores and surfaces to trap sediment and suspended particles.

- The ability of a mechanical filtration system to remove particulate waste is dependent on the pore size of the filter(s).

- *Xenopus* modular aquaria are designed with a pre-filter system to collect the largest particles and a finishing filter that collects smaller sediment.

- As the filters fill with debris, the ability to function is reduced. A consistent action log for checking, cleaning, and replacing these filters is critical for optimal performance.

Biological Filtration

- The biological filter is the most effective way of controlling waste ammonia (Whitaker 2001). The biological filter offers a substrate for growth and maintenance of the naturally occurring bacteria (*Nitromonas* spp. and *Nitrobacter* spp.) responsible for *nitrification*—the breakdown of ammonia and nitrogenous waste products from dissolving organic material, uneaten food, and *Xenopus* waste. The process, known as the **nitrogen cycle**, is shown in **Figure 15**.

- In laboratory *Xenopus* aquaria, the nitrogen cycle lacks the plant life that is an important component of nitrification and the conversion of ammonia and nitrogenous waste into harmless by-products. Instead, buildup of ammonia, nitrite, and nitrate is managed by frequent water replacement.

- In static tanks, ammonia, nitrite, and nitrate buildup is managed by replacing 50–85% or more of the water every day, or several times a week. In flow-through tanks, nitrate and nitrite levels are managed by a slow but constant and high-volume water change. In a recirculating system, a partial water change (usually ~10% of the total water in the system is changed per day) prevents nitrate and nitrite buildup, depending on the number of frogs housed in the system (greater stocking densities mean more frequent changing or greater volumes of water must be changed).

- *Nitromonas* and *Nitrobacter* bacteria must be present and healthy to provide the optimal aquatic environment. They grow attached to various surfaces in the tank. In modular housing systems, nitrifying bacteria grow on submerged

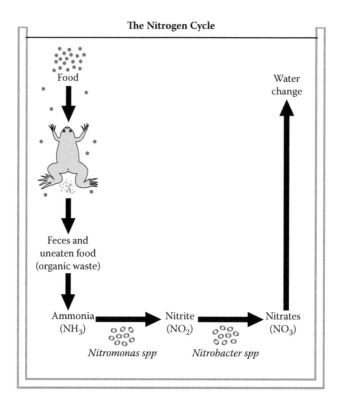

Figure 15 The nitrogen cycle of a *Xenopus* tank.

media surfaces such as plastic or ceramic material shaped as small beads, rings, spheres, or saddles.

- *Nitromonas* bacteria convert waste ammonia (NH_3) to nitrite (NH_4^+), which is toxic. *Nitrobacter* then convert the nitrite (NH_4^+) to the much less toxic nitrate (NO_3). However, ammonia and nitrite can accumulate and reach toxic levels if the frog-stocking density exceeds the conversion capacity of the biological filter (usually when large numbers of animals are added to the tank at once), if there is a mechanical failure in the system (usually a filtration problem or a problem with water changing), or if drugs or chemicals that are lethal to the biological filter are added to the water (during a disease outbreak). Upon failure of the biological filter, toxic ammonia will accumulate and result in sick frogs.

- All surfaces that come into contact with system water acts as a biofilter—for example, a substrate upon which bacteria can affix themselves. *Xenopus* aquaria rack systems typically use passive filters such as ceramic media beads.

- Biological filters are living entities and dependent on stable water quality, including water pH, temperature, and dissolved oxygen (DO). Biofilters take time to grow and mature. Abrupt introduction of large numbers of *Xenopus* into a new modular rack system may result in overwhelming the biofilter with large amounts of waste. Typically, the biological filter takes 4 to 8 weeks to acclimate after the introduction of a few frogs and become established enough such that ammonia and nitrogen are efficiently oxidized into nitrite and nitrate and a 10% water exchange per day is sufficient to keep the nitrate concentration in check. Once the nitrite–nitrogen concentration ratio falls to <0.1 mg/L, the biofilter can be considered acclimated.

- Commercial preparations of cultured nitrifying bacteria— such products as Bio-Spira, Marine® (United Pet Group, division of Spectrum Brands, Inc., Atlanta, GA), and Cycle® (Hagen, Mansfield, IL)—can be used to boost or assist with establishing the biological filter in a new tank or in a tank in which the biological filter has been compromised.

- The biological filter will die in frog tanks that are allowed to stand empty over a period of several weeks. The tanks must be reacclimated by gradually adding just a few frogs before large numbers of new frogs are introduced.

Chemical Filtration

- Chemical filters remove dissolved waste from the water.

- The most common type of chemical filtration is via filtering water through granual activated carbon (GAC). Microspores in the GAC trap waste, including heavy metals and organic molecules by adsorption and ion exchange.

- Chemical additives to the water that can bind and neutralize toxic ammonia, chlorine/chloramines, and nitrite can assist the chemical filters. Examples of such chemical treatments for aquaria water include Amquel®, Instachlor®, AquaSafe® (Tetra Holding, US Inc., Blacksburg, VA), and NovaAqua® (Kordon LLC, Division of Novalet TIC, Hayward, CA).

Ultraviolet Sterilization

- UV sterilization is a disinfecting mechanism used largely in modular recirculating systems to disinfect and clarify the water.

- UV sterilization limits the growth of fungi, viruses, nematodes, and protozoa. Water is passed over an ultraviolet bulb that is protected from direct exposure to the waste by a quartz glass sleeve.

- The UV kill rates for aquatic microbial pathogens are reported as the microwatt seconds per square centimeter (micrW-sec/cm^2), or the power of the light in relation to the time it is exposed to a given area of water and the surface area of that section of water. Selecting a UV light wave and bulb kill rate specific to the pathogen(s) of concern is important to the particular housing system.

- UV kill rates for common aquatic pathogens are shown in **Table 3**.

- The effectiveness of the UV systems against a potential pathogen is determined by the wattage of the bulb, the flow rate, the contact time of water passing over the bulb, the clarity of the water, and the cleanliness of the quartz sleeve (biofilm builds up on it over time). As a general rule, a 2-log reduction in bacteria is the recommended target for tank water samples collected after passing through the UV system. If this is not achieved after the water has passed through the UV system, consider changing the UV bulbs and cleaning the water and tanks to remove debris (see below).

- Excessive debris or particles suspended in the water protect the pathogens from UV; therefore, proper mechanical filtration and prompt replacement of dirty or burned-out UV bulbs are critical to the sanitation of the system.

- UV bulbs have a limited life span (6–9 months), are expensive and can be easily damaged by handling. Gloves should be worn to prevent fingerprints on the bulbs and interference with the UV light rays. A consistent action log for changing and scheduled maintenance should be in place. Most UV light bulbs designed for use with a recirculating rack-style aquarium must be replaced every 9 months.

TABLE 3 UVA KILL RATES FOR COMMON AQUATIC PATHOGENS[a]

Microorganism	pWs/cm²
Aspergillus niger (mold)	330,000
Bacillus subtilis (spores)	22,000
Chlorella vulgaris (algae)	22,000
Clostridium tetani	23,100
Dysentery bacilli	4,200
E. coli (coliforms)	6,600
Fungi	45,000
Ichthyophthirius sp. (ich), tomite	336,000
Influenza (virus)	6,800
Mycobacterium tuberculosis	10,000
Nematode eggs	92,000
Paramecium (protozoa)	200,000
Penicillium digitatum (mold)	88,000
Pseudomonas aerugenosa	10,500
Saccharomyces spp. (yeast)	17,600
Salmonella spp.	15,200
Salmonella typhimurium	15,200
Saprolegnia spp. (egg fungus), zoospore	35,000
Sarcina lutea	26,400
Shigella paradysenteriae	3,400
Staphylococcus aureus	6,600
Streptococcus lactis	8,800
Tobacco mosaic (virus)	440,000
Trichodina nigra	159,000
Tricodina spp.	35,000
Yeast cells (brewer's)	6,600

[a] μWs/cm² = exposure to ultraviolet light of 253.7 nm wavelength in microwatt seconds per square centimeter.

Source: Adapted from http://www.cleanwaterstore.com/mighty-pure-uv.html.

water quality

Standardized water quality values have not been established for *Xenopus*, but parameter ranges considered safe for fish are generally considered safe for *Xenopus*. *ILAR* has published recommendations regarding the parameters that should be monitored for

TABLE 4 GENERAL WATER QUALITY GUIDELINES
FOR LABORATORY XENOPUS[a]

Parameter	X. laevis	X. tropicalis
Temperature (°C)	17–24	24–25
Total Dissolved Gases (% Saturation)	<10	<10
Dissolved Oxygen (mg/L)	>7	>5
pH	6.5–8.5[b]	6.5–8.5[b]
Conductivity (μS)	500–3000	500–1000
Total Hardness (mg/L as $CaCO_3$)	175–300	100–300
Total Alkalinity (mg/L as $CaCO_3$)	50–200	50–200
Toxicity		
Total Ammonia Nitrogen (mg/L)	<0.5[c]	<0.5[c]
Unionized Ammonia (NH_3 –N) (mg/L)	<0.02	<0.02
Nitrite NO_2–N (mg/L)	<0.5	<0.5
Nitrate NO_3–N (mg/L)	<50	<50
Total Chlorine (mg/L)	0.00	0.00
Copper (mg/L)	0.00	0.00

[a] Table provided courtesy of Eric Herbst with modifications by S. L. Green.
[b] Nitrification efficiency is reduced at lower pH. In recirculating systems a minimum pH of 7.0 is advisable.
[c] Value may be higher if pH and temperature bring unionized ammonia concentration below 0.02 mg/L.

freshwater fish and amphibians: pH, temperature, dissolved O_2, conductivity, chlorine, ammonia, nitrite, nitrate, and heavy metals (DeTolla et al. 1995). Acceptable water quality ranges for each of these parameters are reviewed below and shown in **Table 4**. Important differences between *Xenopus laevis* and *Xenopus tropicalis* with regards to water quality are highlighted in the table and discussed below.

pH

The pH is the measure of the ratio of hydrogen ions (H+) to hydroxyl ions (–OH). Many biological processes involve chemical reactions involving the highly reactive hydrogen ions. The pH scale (0 to 14) is derived from the negative logarithm in base 10 of the hydrogen ion concentration. The lower the pH value, the higher the concentration of H+. Hydrogen ions increase acidity; hydroxyl ions increase alkalinity. A pH of 7 is considered neutral. Below 7 is acidic and above 7 is basic. A difference of 1 unit on the pH scale indicates a 10-fold change in the concentration of H+.

- *Xenopus* prefer water pH to be slightly alkaline: pH 7.4–7.5. However, they are highly adaptable and can thrive in water with a pH ranging from 6.0 to 9.0. Rapid changes in pH can result in stress and mortality, especially of *Xenopus* embryos. In general, pH changes should not exceed 0.2 pH units per hour or more than 1.0 pH unit per 24 hours.

- The pH affects the two forms of ammonia found in water: ionized and unionized ammonia. Water with a high pH results in a higher fraction of unionized ammonia (which is more toxic to *Xenopus*) than ionized ammonium (more prevalent at lower pH). Organic matter, respiration, and filtration processes increase acidity (lower the pH). Water pH values <7.0 can inhibit nitrifying bacteria and the efficiency of the nitrogen cycle.

- The pH can decline and ammonia accumulate as a result of increased stocking density, excessive feeding and accumulation of animal waste, damage to the biological filter, or mechanical failure of the filters.

- The pH can be measured by pH test strips or pH meters (pH meters are more accurate). Sodium bicarbonate or calcium carbonate in the form of crushed coral or aragonite can be added to raise or buffer pH levels. Glacial acetic acid or hydrochloric acid can be added to decrease the pH. Altering pH in the presence of animals is risky. Automated monitoring and dosing systems are an advantage of recirculating housing systems and help maintain a stable pH.

Alkalinity

- Alkalinity is the measure of the buffering capacity of water, that is, the ability to resist the change in pH if an acid is added.

- Bicarbonates, carbonates, hydroxides, borates, phosphates and silicates in the water contribute to alkalinity and the water-buffering capacity. Alkalinity in aquaria water is usually expressed in equivalents of calcium carbonate ($CaCO_3$), a major base with good buffering capacity and the ability to neutralize hydrogen ions.

- Alkalinity in a *Xenopus* housing system should not drop below 50 mg/L (as $CaCO_3$) because it is critical to ensure nitrification. Alkalinity buffers or slows the rate of pH changes and provides a carbon source for the nitrifying bacteria that oxidize ammonia to nitrate. This process produces acid, and alkalinity is consumed. Alkalinity and the pH of the water in a *Xenopus* housing system, especially a recirculating system, will generally decrease over time.

- When the pH and alkalinity drop, sodium bicarbonate can be added to increase the alkalinity and pH to the desired level. Ion exchangers or dilution with purified water can be used to reduce the alkalinity of water.

- Alkalinity can be measured with titration test kits.

Temperature

- *Xenopus laevis* typically prefer warm water (21° to 22°C); however, healthy populations can also be found around the world at temperatures ranging from 13° to 30°C (Tinsley et al. 1996). In contrast, wild *X. tropicalis*, largely confined to the lowland rain forest of West Africa, prefer much warmer water temperatures, from 25° to 30°C. Most laboratories keep *X. tropicalis* froglets and adults in water temperatures ranging between 24° and 25°C. *Xenopus tropicalis* do not thrive in temperatures below 22°C and might be more susceptible to disease.

- Temperatures > 30°C are lethal to *X. laevis* and *X. tropicalis*.

- *Most laboratory Xenopus laevis* are housed in water temperatures ranging from 21° to 22°C, and laboratory *X. tropicalis* are typically maintained in water temperature at 24° ± 1°C.

- In a 1998 survey, most facilities reported keeping *X. laevis* at 21° ± 1°C, and most reported having to actively heat the water with thermostatic heaters to maintain constant temperatures (Major and Wassersug 1998). It is important not to place heaters directly into the tank where the frogs are housed as the animals tend to congregate around and beneath submerged heaters and can be burned.

- Ambient temperature can affect the water temperature in frog housing tanks, causing variations of several degrees depending on the type of housing, the room size, ventilation, and the number of heat-producing sources in the room (lights and motors for water chillers/heaters, for example). Water temperature should therefore be monitored constantly and heated or cooled accordingly.

- If the water temperature is suboptimal (too cold), frogs will eat less and become less active (Hutchison and Keith Dupré 1992). Their metabolism and immune systems will be depressed. Oocyte growth, ovulation, spawning, and breeding and growth rates increase with increasing water temperature.

- *Xenopus* are sensitive to temperature changes within the room as well as the temperature changes of water. Water stored for cage changing must be kept constant within 1° to 2°C of the regular tank water, since greater variations can induce thermal or cold shock and sudden death in frogs (Green et al. 2003).

- Gradual increases in water temperature are tolerated, but abrupt changes (> 2° to 5°C) are not and will result in mortality due to heat shock (Green et al. 2003).

- Water temperature has a profound effect on water quality parameters. For example, dissolved oxygen levels decrease as water temperature increases. Unionized ammonia levels increase as water temperature increases (**Table 5**).

TABLE 5 EFFECT OF WATER TEMPERATURE ON WATER UNIONIZED AMMONIA AND pH

Temp °C	6.0	6.5	7.0	7.5	8.0	8.5	9.0
12	0.0218	0.0688	0.217	0.684	2.13	6.44	17.0
13	0.0235	0.0743	0.235	0.738	2.30	6.92	19.0
14	0.0254	0.0802	0.253	0.796	2.48	7.43	20.2
15	0.0274	0.0865	0.273	0.859	2.67	7.97	21.5
16	0.0295	0.0933	0.294	0.925	2.87	8.54	22.8
17	0.0318	0.101	0.317	0.996	3.08	9.14	24.1
18	0.0343	0.108	0.342	1.07	3.31	9.78	25.5
19	0.0369	0.117	0.368	1.15	3.56	10.5	27.0
20	0.0397	0.125	0.369	1.24	3.82	11.2	28.4
21	0.0427	0.135	0.425	1.33	4.10	11.9	29.9
22	0.0459	0.145	0.457	1.43	4.39	12.7	31.5
23	0.0493	0.156	0.491	1.54	4.70	13.5	33.0
24	0.0530	0.167	0.527	1.65	5.03	14.4	34.6
25	0.0569	0.180	0.566	1.77	5.38	15.3	36.3
26	0.0610	0.193	0.607	1.89	5.75	16.2	37.9
27	0.0654	0.207	0.651	2.03	6.15	17.2	39.6
28	0.0701	0.221	0.697	2.17	6.56	18.2	41.2
29	0.0752	0.237	0.747	2.32	7.00	19.2	42.9
30	0.0805	0.254	0.799	2.48	7.46	20.3	44.6

Source: Adapted from Emerson et al., 1975.

Conductivity

- Conductivity is the ability of water to carry an electrical charge and is indicative of the amount of electrolytes, ions, and minerals in solutions. Freshwater conductivity is measured in microsiemens (μS) and is used as an indicator of the efficacy of reverse osmosis and other water purification systems. The conductivity of deionized or reverse-osmosis-treated water will be near 0 μS.

- Conductivity of purified water must be increased in frog housing systems by adding the minerals, trace elements, and ions important to frog osmoregulation. Artificial sea salts are usually used for this purpose. Artificial sea salts (products such as Instant Ocean®, Bio-Sea Marinemix®, and Coralife® salts) contain the proper salt balance and can also be used to increase water alkalinity and hardness.

- Conductivity is measured with a meter and can be affected by temperature, organic materials, and salts and residues.

- Laboratory *Xenopus laevis* are adaptable to a wide range of water conductivity and are usually housed in water with a conductivity of 300 to 2000 μS. *X. tropicalis* are typically housed in water with a conductivity of 500 to 1000 μS.

Hardness

- Water hardness is a measure of the concentration of the dominant divalent cations (ions with a 2+ charge): calcium and magnesium. Hardness is often expressed in terms of the concentration of calcium carbonate ($CaCO_3$) or as degrees of hardness (dH). One dH is equal to 17.9 mg/L hardness as $CaCO_3$. Total hardness as $CaCO_3$ of most natural waters is similar to total alkalinity as $CaCO_3$ because calcium carbonate and magnesium are the primary contributors to alkalinity.

- Total water hardness ranges from 0 to 75 mg/L for soft water to 300 mg/L $CaCO_3$ for very hard water. Most laboratory *Xenopus* are housed in water with a hardness ranging between 100 and 300 mg/L $CaCO_3$.

- Adult female *Xenopus* appear to prefer hard water (Godfrey and Sanders 2004) for optimal egg production. Increasing the general water hardness (general water hardness is based on the concentration of calcium and magnesium ions rather than carbonates and bicarbonates) to >175 ppm appears to increase the firmness of embryos for microinjections.

- Hardness is measured by test kits that use titration.

Ammonia (NH₃)

- In aquatic environments, ammonia is the product of decaying organic matter (predominantly uneaten food in *Xenopus* housing systems) and animal waste (feces, urine).

- Total water ammonia nitrogen is of two forms: ionized ammonia (NH_4^+), which is not highly toxic to *Xenopus*, and unionized ammonia (NH_3), which is toxic.

- The proportion of NH_4^+ and NH_3 is directly related to the water pH, temperature, and salinity/conductivity. NH_3 increases when pH and temperature increase and when salinity/conductivity decreases. Low pH and low water temperature reduce ammonia to the far less toxic ammonium.

- Total ammonia nitrogen can be measured with colorimetric tests. See http://www.aquanic.org/images/tools/ammonia.htm for a calculator that can assist with determining ammonia levels across a wide range of water pH, temperatures, and conductivity.

- *Xenopus* can tolerate much higher levels of ammonia than can fish; however, high ammonia levels in the water are stressful, can inhibit ammonia excretion, and can cause *Xenopus* to switch to ureotelic excretion (an excretion method that requires more energy). High ammonia levels will also adversely affect developing *Xenopus* embryos.

- Total ammonia levels should be maintained below 0.50 mg/L, and NH_3 should be kept below 0.02 mg/L.

- Elevated water ammonia levels in *Xenopus* housing systems can be managed by water changes and by ensuring a healthy biological filter. Chemicals that neutralize ammonia—Pro-Ammonia Detox® (Kent Marine, Franklin, WI), AmQuel®, Ammo Lock® (Mars Fishcare Inc., Chalfont, PA), AmGuard™ (Seachem Laboratories, Madison, GA), Ammonia DeTox™ (Tetra Werke, Melle, Germany)—can be used to assist.

Nitrate and Nitrite

- As shown in **Figure 15**, nitrate and nitrite are the products of bacteria and are part of the nitrogen cycle in aquaria.

- Nitrite (NO_2^-) is the intermediate product of nitrification resulting from the oxidation of ammonia (NH_3) by *Nitromonas* spp. bacteria.

- Nitrite can be absorbed through the skin of *Xenopus* and is highly toxic. Nitrite will oxidize hemoglobin into methemoglobin, thus interfering with the oxygen-carrying capacity.

- Ideally, NO_2^- will be rapidly oxidized by *Nitrobacter* spp. bacteria to nitrate. Increases in NO_2^- occur when there is disruption of the mechanics of the housing system or the biological filter is damaged by the addition of drugs or antibiotics. Nitrite should be kept below 0.5 mg/L and can be kept in check by frequent water changes and ensuring a healthy biological filter.

- Nitrite and nitrate can be measured with colorimetric tests.

- Nitrate (NO_3^-) is the end product of nitrification and results from the oxidation of nitrite by *Nitrobacter*. In natural water systems (ponds, lakes, etc.) NO_3^- is used by plants as an essential nutrient and is also converted to nitrogen gas by anaerobic bacteria. In recirculating housing systems, the concentration of NO_3^- builds up before reaching equilibrium with feeding and water exchange rates. The maximum concentration of 50 mg/L NO_3^- is the general upper limit with no observed adverse effect level (NOAEL) for *X. laevis* tadpoles.

- Periodic water changes will avert the accumulation of nitrate.

Chlorine/Chloramines

- Chlorine and chloramines (reported on water quality tests as "total chlorines") are chemicals added to potable water sources to keep bacteria counts within safe ranges for human consumption.

- Many municipalities now use the odorless, tasteless chloramines—and not chlorine—in tap water. Both chloramines and chlorine are extremely toxic to aquatic species.

- Chloramines are especially dangerous because they are more stable than chlorine, and because they are colorless, tasteless, and odorless.

- Chloramines do not evaporate from water surfaces as readily as chlorine; therefore, the levels will not rapidly decline in standing water. Thus, allowing chloraminated potable water to stand 24 to 36 hours before use in frog housing tanks is not effective and should not be practiced. Chloramine must be removed by special catalytic carbon-activated filters (see Chapter 2). Chlorine/chloramine neutralizing chemicals such as sodium thiosulfate or Chlorine Heavy Metal Neutralizer (Pond Care, Aquarium Pharmaceuticals, Chalfont, PA) can be added to the tank water to assist.

- Both chlorine and chloramines will kill the microorganisms that provide the biological filter, and they will damage the protective slime coat of the frogs.

- Total chlorine can be measured with colorimetric disposable kits found in pet stores or with more specialized spectrophotometric tests.

- Most potable tap water has lethal levels of chlorine or chloramines (0.20–0.80 mg/L), which should be removed prior to using the water for *Xenopus* aquaria. There should be no chlorine or chloramines in the water used for *Xenopus* housing.

Dissolved Oxygen

- Air contains approximately 21% oxygen. Oxygen is not readily soluble in water, and the amount of oxygen dissolved in water (DO) is variable: the solubility of oxygen in water is inversely proportional to temperature and salinity, and directly proportional to barometric pressure. *DO decreases with increased salinity.* For example, at 28°C, DO saturation in freshwater at 10,000 ft is ~5.4 mg/L. At 28°C, DO saturation in full strength seawater at sea level is 6.4 mg/L. *DO also decreases with increased water depth.* Sick or compromised frogs will often hang at the water surface where DO is higher. Frogs affected by gas bubble disease (due to supersaturation of the water with air) will seek the deepest parts of the tank where the DO is less (see Chapter 4).

- During periods of high DO in the water, up to 20% of *Xenopus*'s oxygen demand can be met through cutaneous exchange. *X. laevis* tadpoles rely on as much as 50% of their oxygen demand from DO.

- Because adult *Xenopus* are primarily lung breathers and rise to the water's surface to breathe, water oxygen levels ranging from 5 to 7 mg/L are suitable.
- DO is also important to nitrifying bacteria in the water. As much as 2 mg/L of DO can be removed by a pass through the biological filter.
- DO can be measured by test kits using the Winkler method (for review see www.ecy.wa.gov/programs/wq/plants/management/joysmanual/4oxygen.html or http://supercritical.civil.ubc.ca/~blaval/Courses/CIVL545/CIVL545_3_WinklerMethod.pdf), or with DO meters. In tanks with low DO, aerators can be added and solid waste materials should be removed before they are broken down to increase the biochemical oxygen demand.

Total Dissolved Gases—Percent Saturation (TDG%)

- Total gas pressure is defined as the sum of the individual partial pressures of all gases dissolved in water plus the water vapor pressure.
- TDG% is a measurement comparing the total gas pressure of water to the local barometric pressure (the total of the partial pressures of all gasses in the air). Supersaturation of water with gasses can occur when the total gas pressure in the water exceeds the barometric pressure (see Chapter 4).
- For *Xenopus* housing systems, the TDG% should not exceed 102%. Mortality due to gas bubble disease has been observed in laboratory *Xenopus* when the TDG% reached 103% (Green, S.L., unpublished observations).
- Total dissolved gases can be measured with saturometers and TDG can be monitored with systems supplied by commercial aquaculture vendors.

Carbon Dioxide

- Carbon dioxide makes up only 0.035% of air, but it is highly soluble in water.
- Carbon dioxide in the water is produced by frogs through respiration (a significant amount of CO_2 is exchanged through the skin of *Xenopus*) and by heterotrophic bacteria.

- In water, CO_2 forms carbonic acid and will thus lower water pH.
- Carbon dioxide levels below 5 mg/L are typical of healthy aquaria housing *Xenopus*.

Water Clarity

- Clarity can be misleading. There can be many unseen impurities (such as NH_3 or chloramines) in clear, odorless water.
- Stagnant, murky water can also be relatively free of harmful pollutants, particularly if the water contains abundant plant life.
- For captive *Xenopus*, water clarity is an indicator of fouling by feces and food, and clarity may or may not reflect dangerous levels of nitrite, ammonia, chlorine/chloramines, or other toxicants.

Miscellaneous Water Toxicants

- *Xenopus* are fully aquatic and have skin that is permeable to a number of foreign water toxicants, including disinfectants used in cage washers and chlorine/chloramines.
- The presence of heavy metals (which may leach from water pipes, particularly copper) in the water of captive *Xenopus* may have adverse developmental and reproductive effects, such as decreased egg masses and egg viability and increased prevalence of teratogenesis (Fort et al. 2000).
- In addition, there are anecdotal reports that hand lotions and sanitizers, perfumes, and colognes can adversely affect *Xenopus* during handling.
- Aerosolization of disinfectants used to clean floors in aquatics rooms can also contaminate *Xenopus* water. Cleaning substances should not be used in an aquatics room in a manner that promotes misting or aerosolization (high water-hose pressures and heat).

Monitoring Water Quality

- Densely stocked tanks should be monitored daily, as should problem tanks (where frogs are sick or where water quality has been a historical issue). In the face of disease outbreaks

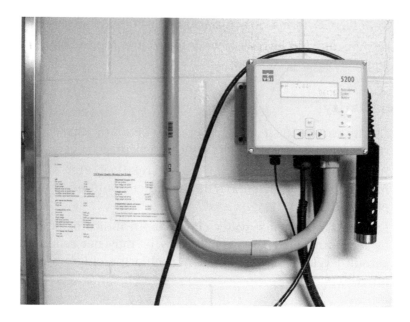

Figure 16 Digital display, automatic water quality monitoring system.

or water quality problems, more frequent monitoring may be required. Water quality probes and monitoring systems with a continuous digital display are most useful, though they are expensive (**Figure 16**).

- Trends and deviations from the recommended range for temperature, pH, ammonia, conductivity, and DO can trigger flashing or audible alarms that alert caretakers and allow immediate action to be taken.

- Commercially available bench-top water test kits such as the Hach® Spectrophotometeric/Colormetric Water Quality test kit system (**Figure 17**) are simple and easy to use, though they are expensive and results may not be ready for an hour or more.

- "Dipstick"-style quick test color strips sold for monitoring fish aquarium water (**Figure 18**) can be used and are convenient, but they are not reliable and should not be the only test used to monitor the water quality of laboratory *Xenopus*.

- Facilities housing aquatics should have an emergency response plan to address the management of water quality in the event of a natural disaster or unexpected loss of power, such that the filters, water heater/chillers, dosers, etc. become

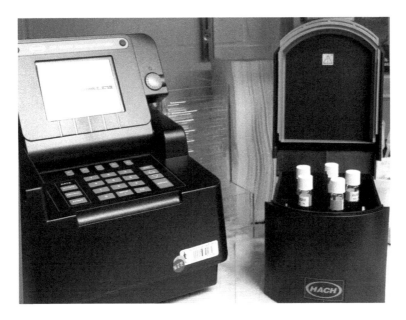

Figure 17 Spectrophotometric/colorimetric water quality test system.

inoperable. Alternate sources of power, water, and housing locations should be identified, along with contact information for the persons responsible for the aquatics systems. An example of an aquatics emergency response plan template is shown in **Table 6**.

stocking density

- To date, large-scale studies on *Xenopus* have not yet been performed to determine the best stocking density for optimal growth rates, health, and fecundity.
- The National Academy of Sciences (1974) recommends 2 L per frog, and a published animal facility survey reports volumes ranging from 1 to 5 L per frog (Major and Wassersug 1998). The same stocking density is also recommended for *X. tropicalis*, even though the species is much smaller than *X. laevis*.
- Two to 4 gallons of water per frog appears to be adequate for *Xenopus* housed in research animal facilities, as long as the animals are able to completely submerge and be adequately covered (Wu and Gerhart 1991).

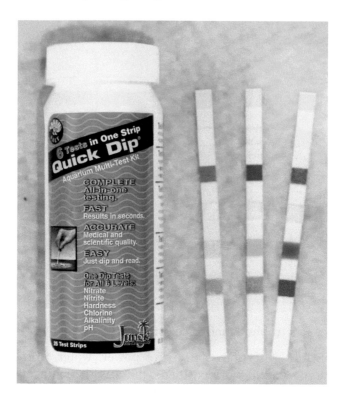

Figure 18 Quick test, color dipstick disposable water quality test kit.

photoperiod

- There is no specific information on the effect of photoperiod on captive *Xenopus*. However, other laboratory amphibian species housed under constant lighting do not produce healthy eggs (Culley 1991).

- Bellerby and Hogben (1938) suggest that photoperiod does not play a significant role in stimulating oogenesis in *Xenopus laevis*, since they normally live year round in darkened water and reproduce over a range of diel cycles.

- For practical purposes, a 12-hours-on/12-hours-off light cycle appears to be appropriate for laboratory *Xenopus*. Light supplied should provide a full wavelength spectrum and allow for visualization of the frogs on all levels of modular racks (tanks on the bottom racks of modules tend to be dark) or additional direct tank lighting provided as needed.

**TABLE 6 AN EXAMPLE EMERGENCY RESPONSE PLAN
FOR AQUATIC AMPHIBIANS**

DATE: _____

AQUATIC AMPHIBIAN
EMERGENCY
RESPONSE PLAN

INSTRUCTIONS:
Post a copy in a prominent location in facility near telephone.

NAME OF FACILITY   ADMINISTRATOR OF FACILITY

FACILITY ADDRESS   TELEPHONE NUMBER
Number   Street   City   State   Zip Code   (   )

I. **Assignments During an Emergency
    (Use Reverse Side if Additional Space Is Required)**

NAME(S)	TITLE	ASSIGNMENT
1.		
2.		
3.		
4		
5.		

II. **Emergency Names and Telephone Numbers**

III. **Facility Exit Locations
      (Using a Copy of the Facility Sketch, Indicate Exits by Number)**

1.	2.
3.	4

IV. **Temporary Relocation Site(s) for Frogs (if Available, Submit Letter
    of Permission from Renter/Lessor/Manager/Property Owner)**

NAME	ADDRESS	TELEPHONE NUMBER ( )
NAME	ADDRESS	TELEPHONE NUMBER ( )

(continued on next page)

TABLE 6 (CONTINUED) AN EXAMPLE EMERGENCY RESPONSE PLAN
FOR AQUATIC AMPHIBIANS

V. Utility Shut-Off Locations (Indicate Location(s) on the Facility Sketch [LIC 999])

ELECTRICITY

WATER

GAS

VI. First Aid Kit (Location)

VII. Equipment

SMOKE DETECTOR LOCATION (IF REQUIRED)

FIRE EXTINGUISHER LOCATION (IF REQUIRED)

TYPE OF FIRE ALARM SOUNDING DEVICE (IF REQUIRED)

LOCATION OF DEVICE

VIII. SPECIAL INSTRUCTIONS

IN THE EVENT WATER FILTERING MECHANICS, THE SYSTEM HEATER OR
CHILLER, OR WATER MAKING UNIT ARE INTERRUPTED DUE TO MECHANICAL
FAILURE OR A POWER OUTAGE, TURN OFF ALL POWER TO THE SYSTEM AND DO
NOT FEED THE FROGS UNTIL FURTHER NOTICE BY THE VETERINARY STAFF.

IX. ALTERNATIVE WATER SOURCE SUITABLE FOR XENOPUS

LOCATION OF WATER SOURCE

AMOUNT OF WATER AVAILABLE

WATER TEMPERATURE

nutrition

There are few evidence-based studies on which to base recommendations for feeding large numbers of laboratory *Xenopus*. Much of what is known and currently practiced is derived from vendor recommendations or time-tested feeding regimens developed by research laboratories, hobbyists, and zoos.

Types of Food

- *Pelleted Feed:* Most facilities housing large numbers of *Xenopus* feed a dry, pelleted commercial food: Frog Brittle® **(Figure 19)** (NASCO, Fort Atkinson, WI) or Aquamax Grower 500® (formerly called Trout Chow, Ralston Purina, St. Louis, MO) (Major and Wassersug 1998). Pelleted food for *Xenopus* typically comes in different sizes (smaller sizes for younger, smaller frogs; pellets will float or sink accordingly), is formulated at various protein concentrations, and is in a powder form for tadpoles.

Figure 19 Frog Brittle® (NASCO, Fort Atkinson, WI), a popular pelleted, complete food developed specifically for laboratory *Xenopus*.

- Trout are carnivorous. Pelleted food designed for this species should meet the protein needs of *Xenopus*. However, trout chow is higher in fat (12%) compared to *Xenopus* Brittle® (6%). The effect of this fat content on laboratory *Xenopus* is unknown (Wright 2001). Aquamax® (trout chow) is considerably less expensive per pound than *Xenopus* Brittle®.

- *Organ Meat:* There is a long history of feeding organ meats to laboratory *Xenopus*, but residual, uneaten tissue debris may be messy and clog housing drainage systems. Large quantities of meat can also be expensive and inconvenient to store. Chlamydia outbreaks resulting in high mortality in *Xenopus* research colonies have been associated with the feeding of organ meat (Wright 2001). Feeding beef liver alone has been associated with vitamin A toxicity and metabolic disease in amphibians (Wright 2001).

- *Live Prey:* Laboratory *Xenopus* will consume live prey (worms, crickets, etc.) with vigor; however, live prey alone is expensive and not practical for large-scale operations housing adult *X. laevis*. There is concern that live prey may transmit parasites and disease, although this has not been documented. Several large research labs housing *X. tropicalis* feed a combination of frog brittle and California blackworms (*Lumbriculus variegates*) (**Figure 20**) to optimize growth rates and reduce the time it takes for the frogs to reach sexual maturity.

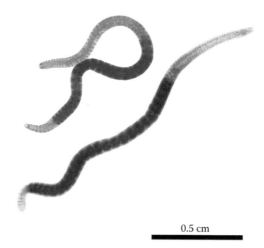

0.5 cm

Figure 20 The California blackworm (*Lumbriculus variegates*) is an important part of the diet for laboratory *X. tropicalis*.

- As do many species, *Xenopus* imprint and develop a preference for certain types of foods. New foods should be mixed with old foods when they are first introduced.

Frequency of Feeding

- Laboratory *Xenopus* should be fed 2 or 3 times per week, although some investigators feed daily (Dawson et al. 1992; Grainger 2008; Major and Wassersug 1998; Sive et al. 2000a).
- The National Academy of Sciences (1974) recommends frequent feeding to decrease the time for *X. laevis* to reach sexual maturity. In the face of water quality problems (emergency power outages, natural disasters, etc.), *Xenopus* can easily go without feed or with infrequent feeding for a week or more. Decreased feeding will help maintain water quality longer in the absence of power for running the filters and UV light sterilization systems.

How Much to Feed

- The exact amount of food to feed per frog or per tank is difficult to determine and depends on the animals' age and gender, the season, and the water temperature.
- The manufacturer of *Xenopus* Brittle® recommends that frogs receive 1 gram of pellets per frog (about 10 pellets per frog).
- Standard resting metabolic rates of anurans (number of kcal required per day by body weight and body temperature) are available (Wright 2001) and can offer a rough guide. For example, a 70-gram frog housed at 20°C should require 0.39 kcal/day.
- The amount of food to feed egg-producing laboratory *Xenopus* is unknown; however, as reviewed by Davis et al. (Davis et al. 2002), commercially reared fish require between 2% and 8% of their body weight in feed in order to grow and maintain oogenesis (less if kept in cold temperatures where metabolic rates are lower). Although such data are lacking for laboratory *Xenopus*, *any* caloric restriction to wild *Xenopus* depletes the energy reserves required for egg production and diminishes growth rates and reproduction (Alexander and Bellerby 1938; Bellerby and Hogben 1938; Duellman and Trueb 1994).
- Frogs are receiving adequate nutrition if after the feeding frenzy the animals have cleared the water of most of the food

(within 15 to 20 minutes). There should be minimal residual food left behind after 2 hours.

- Overfed or stressed *Xenopus*, or *Xenopus* that are handled too soon after eating, will regurgitate their stomach to remove undigested material from the stomach wall mucosa. This might happen without ill effects for the frog, though if the frog is not able to retract its stomach, it will die from suffocation.

- Water quality should be observed after feeding to ensure that the tanks can filter the leftover feed particulate and feces. Feed still floating in the water hours after feeding indicates too much is being fed and the excess should be removed.

sanitation

Sanitation of the components of frog housing systems and related equipment (buckets, nets, and cleaning utensils) can be a challenge because of the sensitivity of these animals to chemical agents. Gross debris should be removed from frog tanks daily by draining or by using vacuums such as those used for cleaning swimming pools.

- The tanks themselves generally require infrequent cleaning (with the exception of tanks that must be depopulated and cleaned due to a disease outbreak), as cleaning will disrupt the biofilm that supports the biological filter and microbial species necessary for a healthy tank. While algae is unsightly, it poses little health risk to the frogs. It can however, become a problem if it is so extensive that it clogs filters and drainage pipes and interferes with the UV light sterilization.

- Some aquaria tanks, depending on their composition, can be cleaned in a standard cage washer system, using water only (no detergents or chemicals).

- Tanks and equipment (nets and buckets) should be cleaned by hand with hot water, dilute, non-alkaline dish soap, and a soft brush or sponge. Stronger disinfectants (harsher detergents, quaternary ammonium, and phenolic compounds) should be avoided because of the sensitivity of *Xenopus* to these chemicals.

- Steam cleaning or cleaning methods that produce aerosols of the water or cleaning agents should be avoided in aquatics rooms. Aerosols can settle over tank water and expose animals to pathogens and harmful chemicals.
- Virkon Aquatics® (Western Chemical, Ferndale, WA) is a mild detergent designed especially for use in aquatic environments. This product is recommended for mopping floors and hand cleaning aquatic tanks and related equipment (do not use Virkon® under high pressure cleaning conditions because of risk of aerosolization). The EPA has restricted the use of Virkon Aquatics® in the state of California.

environmental enrichment

- *The Guide for Care and Use of Laboratory Animals* recommends environmental enrichment for all laboratory animals, including amphibians.
- Amphibians do establish social hierarchies, and although specific studies on captive *Xenopus* are lacking, space-restricted housing into which new individuals are constantly introduced, removed, and reintroduced is the prime environment for dominant conspecifics to establish ruling and exhibit their natural predatory behavior (Hayes et al. 1998). In the absence of refuge cover and an escape route for subordinate animals, they are likely to lose body condition or suffer bite-related fight wounds.
- Clay pots, earthenware pipes, cups, hollow aquarium logs and rocks, polypropylene basket caves, and aquaria lily pads have all been used for refuge cover and enrichment in *Xenopus* aquaria (Brown and Nixon 2004).
- A recent study indicated the introduction of acrylonitrile-butadiene-styrene (ABS) pipes (**Figure 21**) into densely stocked *Xenopus* housing tanks resulted in a dramatic decline in the incidence of bite wounds (Torreilles and Green 2007).
- Refuge cover should be cleaned on a regular basis and should be sinkable, heavy enough that *Xenopus* cannot move the devices around and interfere with tank drainage systems or mechanicals.

Figure 21 Acrylonitrile-butadiene-styrene (ABS) pipe used for environmental enrichment and refuge cover in aquaria housing laboratory *Xenopus laevis.*

identification

- *Xenopus* normally shed and ingest their skin, thus making tattooing or branding ineffective over the long term.

- *Xenopus* can also regenerate their toes, thus toe clipping is not an acceptable method for long-term identification. Toe clipping should not be performed without anesthesia, and animals with an open crush-wound injury after the clipping are at risk for infection by opportunistic pathogens.

- An alternative for identifying individual *Xenopus* is documentation of skin patterns via drawings or photographs. However, patterning or mottling in *Xenopus* can change over time and is dependent on the background coloring of the housing tank, the season, the age of the animal, and other factors.

- Colored plastic or glass beads of appropriate size can be sutured into the toe webs with plastic suture or monofilament suture material (Hoogstraten-Miller and Dunham 1997), or colored leg bands can be applied to the limbs. However, the beads and bands often come off, may be ingested, or may cause constriction.

- Subcutaneous implantation of RFID tagging systems has potential, but has not been tested extensively in *Xenopus* at the time of this publication. Housing large numbers of frogs (in groups of 20 to several hundred or more in a single tank) makes it challenging to track and identify a single RFID-tagged animal. However, the system may prove useful in the identification of breeders. Sources for microchips and directions on their implantation in *Xenopus* can be found in Chapter 6 of this book.

transportation of *Xenopus*

- The International Air Transportation Association publishes *Live Animal Regulations (LAR)* containing guidelines on appropriate shipping container sizes and animal density. The guidelines can be purchased on the IATA website (http://www.iata.org/index.htm).

- *Overcoming the Challenges of Animal Transportation*, published in the Spring 2003 issue of *AAALAC Connections* (view online at http://www.aaalac.org/publications/Connection/Spring_2003_lowres.pdf) reviews regulations and proper documentation that may be required by the USDA and U.S. Customs.

- As stated in Chapter 3, in the United States, shipment permits for *Xenopus* are required by Arizona, California, Hawaii, Montana, Nevada, New Jersey, North Carolina, Oregon, Utah, Virginia, and Washington.

- Regarding **shipping containers**, *Xenopus* can be shipped inside a secondary container that is placed inside a Styrofoam® (Dow Chemical Company, Midland, MI) container that fits snugly inside a cardboard shipping box (plastic containers suitable for food storage, lids must have air holes).

- The frogs should be placed inside a small secondary container and not come into direct contact with the abrasive Styrofoam surface. Lids to secondary containers should be well secured (rubber banded to prevent escape and leakage of wet material).

- Regarding **food and water**, well-soaked (soaked overnight) sphagnum moss (available from nurseries, hobby shops, or hardware stores) or wet sponges should be placed inside the secondary container to prevent the frog from drying out.

Xenopus do not require food during short-term shipments. **Note**: sphagnum moss used for shipping has been associated with mite infestation that caused mortality in a colony of laboratory *Xenopus* (Ford et al., 2004).

- Secondary containers (containing the frogs and the moistened moss or sponges) should be packed inside the Styrofoam box and surrounded with bubble wrap, foam peanuts, or paper packaging to prevent movement.

- Wet or leaking containers can be refused for transport and are a cause of shipment delays.

- Frogs should be shipped when air temperatures do not go below 10°C or above 33°C. During hot weather, cold packs can be placed at the bottom of the Styrofoam container (not in direct contact with the frog). In cold weather, heat packs can be used.

- Live frogs should be shipped by courier overnight or as rapidly as possible. A list of couriers who will accept frog shipments can be found in Chapter 6.

- Frogs in transit should, at a minimum, be observed just prior to departure and closure of the shipping container and immediately upon arrival at the new destination.

- Communication and coordination between the receiving person and shipping persons are critical.

- Shipment and receiving of live frogs should best be scheduled for early during the week. Avoid shipping frogs over the weekend or holidays.

- Upon arrival, newly transported frogs should be immediately examined by a veterinarian or knowledgeable designated caretaker.

record keeping

The following are some of the basic records that should be kept on laboratory *Xenopus*.

- **Medical health records**—Sick frogs should be isolated and a daily observation/treatment sheet maintained.

- **Experimental records**—Experimental manipulations should be documented and perioperative monitoring records maintained.

- **Census**—Census should be estimated or based on shipment information, especially when large numbers of frogs (50 to 100 plus frogs) are kept in the same tank (where it may be too hard to count individual animals). Weekly census is helpful for not only accounting purposes: sporadic mortalities in the population may be the first sign of infectious disease outbreaks or of water quality problems and should be reported to the veterinarian.

- **Work records**—Routine husbandry tasks should be documented as performed: feeding, water quality testing results/ parameters, cleaning and vacuuming of tanks, water temperature, water exchanges, water treatments (addition of conditioners or salts), and filter changes. The date and time the work was performed should be noted. Work records should be initialed by those performing the task.

references

Alexander, S. S. and C. W. Bellerby. 1938. Experimental studies on the sexual cycle of the South African clawed toad (*Xenopus laevis*). I. *J. Exp. Biol.* 15:74–81.

Bellerby, C. W. and L. Hogben. 1938. Experimental studies on the sexual cycle of the South African clawed toad (*Xenopus laevis*). III. *J. Exp. Biol.* 15:91–100.

Brown, M. J. and R. M. Nixon. 2004. Enrichment for a captive environment—the *Xenopus laevis. Anim. Tech. Welfare* 3(2):87–95.

Council of Europe. 2003. Revision of the appendix A of the Convention ETS123–Species specific provisions for amphibians: Background information for the proposals presented by the Group of Experts on Amphibians and Reptiles (Part B).

Culley, D. D. 1991. *Bufo* culture. In *Production of aquatic animals, world animal science C4*, ed. C. E. Nash, 185–205. Vancouver, Canada: Elsevier Science Publishers.

Davis, C. R., M. S. Okihiro, and D. E. Hinton. 2002. Effects of husbandry practices, gender, and normal physiological variation on growth and reproduction of Japanese medaka, *Oryzias latipes. Aquatic Toxic.* 60(3-4):185–201.

Dawson, D. A., T. W. Schultz, and E. C. Shroeder. 1992. Laboratory care and breeding of the African clawed frog. *Lab. Anim.* 21:31–36.

DeTolla, L. J., S. Srinivas, B. R. Whitaker et al. 1995. Guidelines for the care and use of fish in research. *ILAR J.* 37(4):159–73.

Duellman, W. E. and L. Trueb. 1994. Reproductive strategies. In *Biology of amphibians*, ed. W. E. Duellman and L. Trueb, 13–50. Baltimore: Johns Hopkins University Press.

Ford, T. R., D. L. Dillehay, and D. M. Mook. 2004. Cutaneous acariasis in the African clawed frog (*Xenopus laevis*). *Comp. Med.* 54(6):713–17.

Fort, D. J., E. L. Stover, C. M. Lee, and W. J. Adams. 2000. Adverse developmental and reproductive effects of copper deficiency in *Xenopus laevis. Biol. Trace Elem. Res.* 77:159–72.

Godfrey, E. W. and G. E. Sanders. 2004. Effect of water hardness on oocyte quality and embryo development in the African clawed frog (*Xenopus laevis*). *Comp. Med.* 54:170–75.

Grainger, R. 2008. *Xenopus tropicalis* amphibian model for vertebrate development genetics. http://faculty.virginia.edu/xtropicalis/ (accessed April 17, 2009).

Green, S. L., R. C. Moorhead, and D. M. Bouley. 2003. Thermal shock in a colony of South African clawed frogs (*Xenopus laevis*). *Vet. Rec.* 152:336–37.

Hayes, M. P., M. R. Jennings, and J. D. Mellen. 1998. Beyond mammals: Environmental enrichment for amphibians and reptiles. In *Second nature: Environmental enrichment for captive animals*, ed. D. J. Shepherdson, J. D. Mellen, and M. Hutchins, 205–34. Washington and London: Smithsonian Institution Press.

Hilken, G., J. Dimigen, and F. Iglauer. 1995. Growth of *Xenopus laevis* under different laboratory rearing conditions. *Lab. Anim.* 29:152–62.

Hoogstraten-Miller, S. and D. Dunham. 1997. Practical identification methods for African Clawed Frogs (*Xenopus laevis*). *Lab. Anim.* 26(7):36–38.

Hutchison, V. H. and R. Keith Dupré. 1992. Thermoregulation. In *Environmental physiology of the amphibians*, ed. M. E. Feder and W. W. Burggren, 206–49. Chicago: University of Chicago Press.

Koerber, A. S. and J. Kalishman. 2009. Preparing for a semiannual IACUC inspection of a satellite zebrafish (*Danio rerio*) facility. *J. Am. Assoc. Lab. Anim. Sci.* 48:65–75.

Major, N and R. J. Wassersug. 1998. Survey of current techniques in the care and maintenance of the African clawed frog (*Xenopus laevis*). *Contemp. Topics Lab. Anim. Sci.* 37(5):57–60.

National Academy of Sciences. 1974. *Amphibians: Guide for the breeding, care, and management of laboratory animals.* Washington: National Academy of Sciences Publishing Office.

National Research Council. 1996. Committee to Revise the *Guide for the care and use of laboratory animals.* Washington: National Academy Press.

Nieuwkoop, P. D. and J. Faber. 1967. *Normal table of Xenopus laevis (Daudin). A systematical and chronological survey of the development from the fertilized egg till the end of metamorphosis.* Amsterdam: North-Holland Pub. Co.

Sive, H. L., R. M. Grainger, and R. M. Harland. 2000a. *Early development of Xenopus laevis. A laboratory manual.* New York: Cold Spring Harbor Laboratory Press.

Sive, H. L., R. M. Grainger, and R. M. Harland. 2000b. Introduction. In *Early development of Xenopus laevis: a laboratory manual,* ed. H. L. Sive, R. M. Grainger, and R. M. Harland, 1–11. New York: Cold Spring Harbor Laboratory Press.

Smith, L. D., W. L. Xu, and R. L. Varnold. 1991. Oogenesis and oocyte isolation. *Methods Cell Biol.* 36:45–60.

Tinsley, R. C., C. Loumont, and H. R. Kobel. 1996. Geographical distribution and ecology. In *The biology of Xenopus,* ed. R. C. Tinsley and H. R. Kobel, 35–60. New York: Oxford University Press.

Torreilles, S. L. and S. L. Green. 2007. Refuge cover decreases the incidence of bite wounds in laboratory South African clawed frogs (*Xenopus laevis*). *J. Am. Assoc. Lab. Anim. Sci.* 46:33–36.

Whitaker, B. R. 2001. Water quality. In *Amphibian medicine and captive husbandry,* ed. K. M. Wright and B. R. Whitaker, 147–57. Malabar, FL: Krieger Publishing Company.

Wright, K. M. 2001. Diets for captive amphibians. In *Amphibian medicine and captive husbandry,* ed. K. M. Wright and B. R. Whitaker, 63–72. Malabar, FL: Krieger Publishing Company.

Wu, M. and J. Gerhart. 1991. Raising *Xenopus* in the laboratory. *Methods of Cell. Biol.* 36:3–18.

management

---

regulations and regulatory agencies

Xenopus are ectotherms (cold-blooded) and are therefore not subject to oversight and regulations by the United States Department
of Agriculture (USDA) according to the Animal Welfare Act (AWA).
However, other agencies and organizations in the United States have
oversight or accreditation responsibilities for programs that use
Xenopus in teaching, research, or testing.

- All vertebrate research sponsored by the U.S. government must
  comply with the *US Government Principles for the Utilization and
  Care of Vertebrate Animals Used in Testing, Research and Training (PHS Principles)* (Office for Protection from Research Risks
  1986) and is thus subject to the general policies as described
  in the *Guide of the Care and Use of Laboratory Animals* (*The
  Guide*) published by the National Research Council (NRC)
  (1996). These principals and policies apply to research conducted on *Xenopus* by federally funded institutions.

- *The Guide* considers *Xenopus* to be a non-traditional laboratory species. While *The Guide* does not offer specific recommendations for housing and handling, it does recommend
  expert advice be sought regarding the care of these animals
  in laboratory research and field studies.

- Adequate care, housing, and veterinary oversight of this species is also expected to be reviewed by the institution's Animal
  Care and Use Committee (IACUC) and by any institution

desiring accreditation by the Association for Assessment and Accreditation of Laboratory Animal Care International (AAALAC International).

- In the United States, *Xenopus* are a non-native, invasive species, and many states have enacted specific restrictions regarding their use and transport. The United States Department of the Interior (USDI) delegates the oversight of non-native, invasive species in interstate and international commerce to the Fish and Wildlife Services (FWS). FWS policies are described at http://www.fws.gov/invasives/laws.html.

- In the state of California, the Fish and Game Department requires special permits for the importation and housing of laboratory *Xenopus* and for *Xenopus* intended for display in zoo exhibits. The sale of *Xenopus* intended to be resold as pets is also prohibited. Many other states, including Arizona, Hawaii, Montana, Nevada, New Jersey, North Carolina, Oregon, Utah, Virginia, and Washington, impose similar restrictions or require transportation permits.

- The FWS exercises its authority through the CITES agreement (Convention on International Trade in Endangered Species of Flora and Fauna) for international commerce and the Endangered Species Act (ESA) for interstate commerce. For full text of the convention and of the ESA go to http://www.cites.org/eng/disc/text.shtml and to http:www.nmfs.noaa.gov/pr/laws/esa/, respectively.

- Investigators should become familiar with local wildlife laws if they intend to collect specimens or conduct field studies using *Xenopus*.

occupational health and safety: injury and zoonotic risks

In the United States, PHS policy requires institutions to have an occupational health and safety program for individuals working with laboratory animals, including aquatics. Specific elements of an occupational health and safety program are described in *Occupational Health and Safety in the Care and Use of Research Animals* published by the National Research Council (NRC). Principal investigators should ensure their laboratory staff are informed of and participate in their institution's program.

Although the hazards of working specifically with *Xenopus* appear to be small, the most significant risks are:

- *Slipping and falling, physical injury, and shock hazard due to the wet environment.* Rubber non-slip mats are recommended for high-traffic wet areas. Electrical outlets, switches, and appliances should be properly constructed (water proofed and ceiling mounted) and equipped with ground fault interruption devices. Caretakers should wear rubber boots or comparable footwear.

- *Contact with infectious or potentially zoonotic diseases.* Several species of *Mycobacterium* and *Chrysobacterium* (formerly called *Flavobacterium*) and other aquatic opportunistic pathogens have the potential to cause infections in humans and are known to infect laboratory *Xenopus* (Chai et al. 2006; Clothier and Balls 1973; Godfrey et al. 2007; Green et al. 1999; Green et al. 2000; Olson et al. 1992; Schwabacher 1959; Suykerbuyk et al. 2007; Taylor 2001; Trott et al. 2004). These bacteria are ubiquitous and are normally present in tank water. Other potential zoonotic pathogens include *Cryptosporidium* (Green et al. 2003) and *Salmonella* spp. (Anver and Pond 1984). Appropriate waterproof personal protective equipment should be provided.

To date, there are no documented reports of research laboratory workers becoming infected with a potential pathogen as a result of working with laboratory *Xenopus*. However, aquatics personnel (working in the fish aquaculture industry) have become infected with *M. marinum* (a condition also known as "fish TB"), usually through skin abrasions on areas of the body that came into contact with the water (Petrini 2006). Symptoms include a papule, or nodule, and redness and swelling at the abrasion site, and occasionally patients complain of a persistent cough. Mycobacteriosis acquired in this manner is very rare (occurrence rate is ~0.25 cases in 100,000 aquatics workers) and usually not life threatening; however, treatment requires antibiotics (Rallis and Koumantaki-Mathioudaki 2007).

Personnel working with *Xenopus* should wash hands frequently and wear gloves when handling the animals or related equipment and, as recommended above, wear appropriate personal protective equipment.

references

Anver, M. R. and C. I. Pond. 1984. Biology and diseases of amphibians. In *Laboratory animal medicine*, ed. J. G. Fox, B. J. Cohen, and F. M. Loew, 427–47. Orlando, FL: Academic Press, Inc.

Chai, N., L. Deforges, W. Sougakoff et al. 2006. *Mycobacterium szulgai* infection in a captive population of African clawed frogs (*Xenopus tropicalis*). *J. Zoo Wildlife Med.* 37:55–58.

Clothier, R. H. and M. Balls. 1973. Mycobacteria and lymphoreticular tumours in *Xenopus laevis*, the South African clawed toad. I. Isolation, characterization and pathogenicity for *Xenopus* of *M. marinum* isolated from lymphoreticular tumour cells. *Oncology* 28:445–57.

Godfrey, D., H. Williamson, J. Silverman, and P. L. Small. 2007. Newly identified *Mycobacterium* species in a *Xenopus laevis* colony. *Comp. Med.* 57:97–104.

Green, S. L., D. M. Bouley, C. A. Josling, and R. Fayer. 2003. Cryptosporidiosis associated with emaciation and proliferative gastritis in a laboratory South African clawed frog (*Xenopus laevis*). *Comp. Med.* 53(1):81–84.

Green, S. L., D. M. Bouley, R. J. Tolwani et al. 1999. Identification and management of an outbreak of *Flavobacterium meningosepticum* infection in a colony of South African clawed frogs (*Xenopus laevis*). *J. Am. Vet. Med. Assoc.* 214:1833.

Green, S. L., B. D. Lifland, D. M. Bouley et al. 2000. Disease attributed to *Mycobacterium chelonae* in South African clawed frogs (*Xenopus laevis*). *Comp. Med.* 50:675–79.

National Research Council. 1996. *Committee to revise the guide for the care and use of laboratory animals.* Washington: National Academy Press.

Office for Protection from Research Risks. 1986. *Public health service policy on humane care and use of laboratory animals.* Washington: U.S. Department of Health and Human Services.

Olson, M. E., S. Gard, M. Brown, R. Hampton, and D. W. Morck. 1992. *Flavobacterium indologenes* infection in leopard frogs. *J. Am. Vet. Med. Assoc.* 201:1766–70.

Petrini, B. 2006. *Mycobacterium marinum*: ubiquitous agent of waterborne granulomatous skin infections. *Eur. J. Clin. Microbiol. Infect. Dis.* 25:609–13.

Rallis, E. and E. Koumantaki-Mathioudaki. 2007. Treatment of *Myco-bacterium marinum* cutaneous infections. *Expert. Opin. Pharmacother.* 8:2965–78.

Schwabacher, H. 1959. A strain of *Mycobacterium* isolated from skin lesions of a cold-blooded animal, *Xenopus laevis*, and its relation to atypical acid-fast bacilli occurring in man. *J. Hyg. (Lond).* 57:57–67.

Suykerbuyk, P., K. Vleminckx, F. Pasmans et al. 2007. *Mycobacterium liflandii* infection in European colony of *Silurana tropicalis*. *Emerg. Infect. Dis.* 13:743–46.

Taylor, S. K. 2001. Bacterial diseases. In *Amphibian medicine and captive husbandry*, ed. K. M. Wright, and B. R. Whitaker, 159–80. Malabar, FL: Krieger Publishing Company.

Trott, K. A., B. A. Stacy, B. D. Lifland et al. 2004. Characterization of a *Mycobacterium ulcerans*-like infection in a colony of African tropical clawed frogs (*Xenopus tropicalis*). *Comp. Med.* 54:309–17.

veterinary care

physical examination

- The first step to the physical examination of the laboratory *Xenopus* is to observe the animal undisturbed in the tank.

- Sick frogs tend to have buoyancy problems. They float on the water's surface and are reluctant to dive. If profoundly moribund, the front legs relax such that the inside of the front feet turn outward. A tap on the side of the tank will send vibrations through the water that usually result in the healthy frogs diving and darting for cover. Sick frogs may be slow to respond to water disturbances or not respond at all.

- Sick frogs may swim upside down, swim in circles, or swim with a "whole body" tilt to one side or the other.

- An inordinate amount of shed skin in the tank water indicates that a number of frogs are sick and is often indicative of a water quality problem or widespread disease.

- Sick frogs are easier to catch and can be readily restrained for the physical examination.

- Temperature, pulse rate, and respiratory rate are vital signs normally assessed in mammalian patients. However, in aquatic amphibians these parameters are difficult to assess and highly influenced by the water temperature, the frog's age, gender, and metabolic state, and the seasonality.

- Carefully examine the frog's integument for signs of petechia and ecchymosis (easiest to see on the light-colored ventrum), ulcerations, discoloration (redness), changes in texture (from slimy and smooth to roughened), and tufts of cottony fungus.

- Toes should be examined for swelling (a sign of gout) and for injury and loss of a claw. Interdigital webbing should be examined carefully (the use of a magnifying glass or dissecting microscope is helpful) for the presence of engorged blood vessels, petechia and ecchymosis, and gas bubbles (bubbles beneath the skin in interdigital webbing is pathognomonic for gas bubble disease).

- Close inspection of the frog's dorsum under a dissecting microscope or with a magnifying glass may reveal evidence of parasitism (the nematode *Pseudocapillaria*, which tunnels just beneath the slime coat and surface of the skin).

- Eyes should be clear and free of cloudiness.

- Examine the mouth for foreign bodies and for the presence of a regurgitated stomach. The snout should be sharply delineated. Round, blunted snouts are indicative of water retention and osmolality regulation problems.

- Cloacal swelling and reddening can be associated with recent hormonal priming and egg laying, or with septicemia.

- A sunken, hour-glass shape of the coelomic cavity is a sign of weight loss in laboratory *Xenopus*. A large, distended coelomic cavity is indicative of coelomic effusion.

- Palpation of the coelomic cavity may reveal visceral tumors or masses, or an impacted stomach (although it is not uncommon on palpation to feel a stomach filled with a meal in a healthy frog).

quarantine

- The purpose of the quarantine is to allow animals to stabilize following shipment, to seek replacement animals for sick or damaged animals from the vendor, to avoid exposure to incoming disease from new stock, and to condition new animals to current environmental conditions prior to commencement of an experiment.

- All new amphibians should be quarantined on arrival, either by placement in isolated tanks in the regular housing room, or ideally, in a separate aquatics quarantine room.

- The length of time the new frogs are quarantined depends on their source. New arrivals should be quarantined 2 to 3 months before introduction into the rest of the colony, particularly if they are wild-caught frogs. Frogs purchased from vendors that supply only captive-reared and bred frogs may not need to be quarantined as long. If the new animals are healthy, eggs or oocytes can be collected two weeks after the frogs' arrival.

- Maintain separate holding tanks and separate handling equipment (nets, scoops, etc.) for quarantined animals. New arrivals should NOT be placed in a modular housing system where the water is re-circulated and shared with the rest of the established colony.

- Quarantined animals should be examined daily for activity level, skin discoloration, ulceration, petechial (pinpoint) hemorrhages on the legs, abdominal swelling, and any other unusual changes.

- Most adult *Xenopus* are shipped in sealed buckets or boxes with moist foam cubes or damp moss to prevent desiccation. If animals arrive in water, the animals and container should be allowed to adjust to room temperature.

- The shipment water should be kept with the new arrivals and gradually be replaced by slow dilution over several hours with facility water to avoid "shock" to the new animals.

- New arrivals should be handled last. The person providing care should always tend first to established healthy collections and breeding colonies. Caretakers should adhere to frequent hand-washing procedures and glove changes between cages of different shipments and between individual animals.

clinical problems

More is known about the diagnosis and treatment of disease in almost every other vertebrate species than what is known about conditions affecting captive *Xenopus*. However, over the last decade or so, veterinarians have begun to publish more reports on disease outbreaks and the health problems of laboratory *Xenopus*, disseminating a greater knowledge of conditions under which this species gets sick, most of which can be attributed to husbandry, handling, and management practices.

Figure 22 Weight loss (upper frog) in a juvenile female. Note the snout-to-vent lengths are the same in the two frogs and the bite wound on the front leg/axilla of the weight-loss frog.

General signs of illness in *Xenopus* include

- buoyancy problems
- weight loss (**Figure 22**)
- coelomic distention, also referred to in the older amphibian literature as **dropsy**
- whole body "bloat" or **hydrops** (a term also used in older amphibian literature) due to accumulation of fluid in the sub-cutaneous space (**Figure 23 A, B, C, and D**)
- cottony tufts of fungus attached to the skin
- petechia (**Figure 24**), ecchymoses on the belly, legs, nose, or feet.
- excessive slime coat mucous production
- bite wounds on the legs
- cloudy corneas
- hyperemic, distended cloaca
- swimming in circles, tilted to the side, or upside down
- unusual amounts of shed skin in the tank water

Figure 23 Bloat due to fluid accumulation in the loosely attached skin of laboratory *Xenopus*. In A, B, C and D, the accumulation of fluid is in the subcutaneous space. This clinical sign is often called hydrops in the older amphibian literature and is indicative of an osmoregulatory problem often associated with infections or renal disease.

- nonresponsiveness or slow to respond to disturbances in the water
- ulcerative dermal lesions
- exophthalmos
- air bubbles beneath the skin or between the toes in the interdigital webbing

All of these symptoms are signs of ill health, and, with the exception of gas bubbles in the integument, almost none are specific for a particular disease.

Some of the best-described infectious diseases of captive *Xenopus* are reviewed below. The list is not intended to be exhaustive, but these diseases are some of the most frequently reported in laboratory animal medicine. Treatment of infectious disease in laboratory *Xenopus* poses unique challenges and is discussed in greater detail in a separate section.

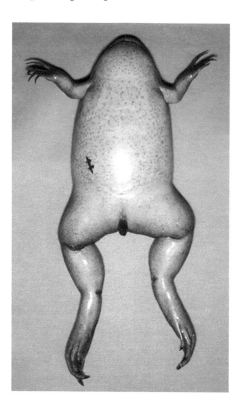

Figure 24 Petechia on the ventrum of an adult laboratory *Xenopus*. This frog was humanely euthanized due to septicemia from *Chryseobacterium* (formerly called *Flavobacterium*). Note the sutures in ventrum after surgical harvest of oocytes. (From Green, S. L. et al., *J. Am. Vet. Med. Assoc.* 214(12): 1833–38, 1999. With permission.)

Bacterial Infections

Red leg syndrome

- "Red leg" is the name attributed to the symptoms of bacterial skin infection: swelling, petechia, and ecchymosis affecting a limb or digit that can progress beyond the local site of infection to systemic disease. Red leg as a symptom is not specific for a single etiological agent but is often associated with bacterial infection.

- Red leg syndrome often occurs as a secondary problem related to physical injury, handling or surgical stress, skin abrasions or erosions and disruption of the slime coat layer, dehydration, underlying systemic infectious disease, parasites, bite wounds, and water quality problems. A single animal may be affected or many frogs in a tank may show symptoms.

- The numerous organisms associated with red leg in captive *Xenopus* are ubiquitous in the aquatic environment and are opportunistic primary or secondary pathogens often found in co-culture with several other pathogens (Densmore and Green 2007).

- *Aeromonas hydrophila*, a motile, Gram-negative rod that produces endotoxins, proteases, and other virulence factors, was the first reported causative agent of red leg in laboratory frogs, in 1905 (Emerson and Norris 1905).

- Infection with other organisms has since also been reported to cause red leg: *Proteus* spp., *E. coli, Aerobacter* spp., *Pseudomonas* spp., *Citrobacter* spp., *Mimi* spp., *Staphylococcus* spp., *Streptococcus* spp. (Crawshaw 1992; Manuel et al. 2002), *Enterobacter* spp., *Klebsiella* spp., and *Chryseobacterium* (formerly called *Flavobacterium)* spp. (Taylor 2001a).

- The clinical signs include lethargy, progressive thickening and swelling of a limb, and the appearance of petechia and ecchymosis, often spreading to the ventrum.

- Diagnosis is by microbial culture of affected tissues, liver, or heart blood.

- Treatment of red leg is often complicated by concurrent fungal skin infection or an underlying systemic disease. Small, localized lesions may improve and slowly heal if the frog is singly housed for 2 weeks in amphibian Ringer's solution (**Table 7**; water changed daily) and Shield-X® (Aquatronics, Malibu, CA) is applied to the area as an aid to the slime coat.

- For the reasons discussed below, systemic treatment with broad-spectrum antibiotics may have limited results and may not be practical when many frogs are involved or when the lesion is advanced. If treatment is attempted, antibiotic therapy should be based on pathogen identification via molecular methods, microbial culture, and antimicrobial sensitivity patterns. **Table 8** lists antibiotics that have been suggested for use in amphibians.

Chryseobacterium (formerly called *Flavobacterium*) spp.

- *Chryseobacterium*, Gram-negative, aerobic rods that are highly resistant to antibiotics and chlorine/chloramines, are reported

TABLE 7 AMPHIBIAN RINGER'S SOLUTION[a]

Distilled Water	1 Liter	1 Gallon
NaCl	6.6 g	25 g
KCl	0.15 g	0.57 g
CaCl$_2$	0.15 g	0.57 g
NaHCO	0.2 g	0.76 g

[a] At 21.7°C, the measured pH of this solution is ~7.99 and the measured osmolality (mmol/kg) is 229. Mix solution to ensure all crystals dissolve. Agitate before use. Keep in closed container to reduce evaporation. Can be used as a bath. Sterilize before systemic (IC or SQ) administration. Can be given IC at a volume not to exceed 2–5% of the amphibian's body weight in a 12-hour period.

Source: Adapted from Kevin M. Wright and Brent R. Whitaker, *Amphibian Medicine and Captive Husbandry*, (Malabar, FL: Krieger Publishing Company, 2001), page 318.

to cause septicemia and death in laboratory *Xenopus laevis* (Green et al. 1999). Specifically, *F. indologenes* and *F. meningosepticum* have been associated with disease in laboratory *Xenopus laevis* (Green et al. 1999). *F. oderans* has also been reported to cause disease in amphibians (Olson et al. 1992).

- *Chryseobacterium* are ubiquitous in aquatic environments and normally inhabit water, soil, raw meat, milk, and other foods and are a normal inhabitant of amphibian and insect intestines. The organisms prefer moist or wet environments and standing water at 21°C and can withstand chlorine.

- Infection with chryseobacteria is thought to be related to stress, surgical manipulation, cool water temperatures (water 21°C and less), shedding of the organism by subclinically infected frogs, or heavy growth in the biofilm of housing tanks (Green et al. 1999).

- Clinical signs are consistent with septicemia: corneal opacity, ascites, subcutaneous edema, inability to dive, lethargy, congestion of the web vessels, petechial hemorrhage over the entire body (see **Figure 24**), and sudden death. In one outbreak at a research animal facility, the mortality rate reached a peak of 35% in a room housing 300 frogs (Green et al. 1999).

- Diagnosis is aided by the detection of toxic neutrophils and Gram-negative intracellular bacteria in white bloods cells on

TABLE 8 SELECTED DRUGS AND COMPOUNDS SUGGESTED FOR USE IN CAPTIVE AMPHIBIANS

Antibiotics

Amikacin	5 mg/kg q 48 h IM for 5–14 tx
Carbenicillin	200 mg/kg SQ, IM, or IC q 24 h
Ceftazadime	20 mg/kg IM q 48–72 h
Chloramphenicol	50 mg/kg SQ, IM, or IC q 24 h; 20 mg/L bath; 10 mg/mL in 0.5% saline as 24 h bath
Ciprofloxacin	10 mg/kg PO q 24–48 h for 7 tx; 25-37.5 mg/gal for 6–8 h bath q 24 h for 7 d
Enrofloxacin	5–10 mg/kg IM q 24 h for minimum 7 tx
Gentamicin	2.5 mg/kg IM q 72 h (cold temp. of 3°C); 3 mg/kg IM q 24 h (warm temp. of 2°C); 8 µg/mL in 0.5% saline as 24 h bath for 5 d
Isoniazid	12.5 mg/L as a 24 h bath
Metronidazole	50 mg/kg PO q 24 h for 3 d, 50 mg/L bath for up to 24 h
Nalidixic acid	10 mg/L bath; 10 µg/ml in 0.5% saline bath for 24 h
Nitrofurantoin	50 µg/ml in 0.5% saline bath for 24 h
Nitrofurazone	10–20 mg/L bath for 24 h; 100 mg/L bath for up to 48 h
Piperacillin	100 mg/kg IM or SQ q 24 h
Tetracycline	50 mg/kg PO q 12 h; 10 ug/ml in 0.5% saline bath for 24 h
Trimethoprim/ sulfamethoxazole	15 mg/kg PO q 24 h up to 21 d; 20 µg/mL in 0.5% saline bath for 24 h; 80 µg/mL in 0.5% saline bath for 24 h
Trimethoprim/ sulfadiazine	15–20 mg/kg IM q 48 h for 5–7 tx
Trimethoprim/sulfa	3 mg/kg SQ q 24 h, 5–7 d

Antifungal Compounds

Acriflavin	500 mg/L for 30 min bath q 24 h
Acriflavin/methylene	0.5 mL of a 0.45% acriflavin/0.00075% methylene blue stock added to 1 L Holtfreter's solution as a 24-h bath for 3–5 d
Amphotericin B1	1 mg/kg IC q 24 h for 14–28 tx
Benzalkonium Chloride	2 mg/L as a 60-min bath q 24 h; 0.25 mg/L, bath for 72 h; 0.25 ppm as a continuous bath
Copper sulfate	500 mg/L for 2 min q 24 h for 5 d, then q 7 d until no longer needed
Itraconazole	2–10 mg/kg PO q 24 h for 14–28 d, 0.01% liquid form in 0.6% saline as a 5-min bath for 11 d
Ketoconazole	10 mg/kg PO q 24 h; 10–20 mg/kg PO q 24 h for 14–28 d
Methylene blue	50 mg/mL as a 10 second dip, 2–4 mg/L as a continuous bath for up to 5 d
Miconazole	5 mg/kg IC q 24 h for 14–28 d
Potassium permanganate	200 mg/L bath q 24 h for 5 min; 1 g/100 ml as topical permanganate tx q 48–72 h

(continued on next page)

TABLE 8 (CONTINUED) SELECTED DRUGS AND COMPOUNDS SUGGESTED FOR USE IN CAPTIVE AMPHIBIANS

Sodium chloride	4–6 g/L bath for 72 h; 10–25 g/L bath for 5–30 min; 20 mg/L bath for 6–8 h

Antiparasitic Compounds—Protozoa

Acriflavin	0.025% bath for 24 h for 5 d
Copper sulfate	500 mg/L dip for 2 min q 24 h; 0.0001 mg/L bath to effect
Methylene blue	2 mg/L as a continuous bath
Metronidazole	10 mg/kg PO once;
	10 mg/kg PO q 24 h for 5–10 d;
	50 mg/kg PO q 24 h for 3–5 d;
	100–150 mg/kg PO q 14–21 d;
	50 mg 1L bath for up to 24 h
Potassium permanganate	7 mg/L bath for 5 min q 24 h
Quinine sulfate	30 mg/L bath for up to 1 h (for hemoprotozoans)
Sodium chloride	4–6 g/L bath for 24–72 h; 6 g/L bath for 3–5 d; 10–25 mg/L bath for 5–30 min
Sulfadiazine	132 mg/kg q 24 h (for coccidiosis)
Sulfamethazine	1 g/L bath (for coccidiosis)

Antiparasitic Compounds—Helminths

Fenbendazole	10 mg/kg PO once; 50 mg/kg PO SID for 3–5 d; 100 mg/kg PO q 10 d, 100 mg/kg PO q 10–14 d, 100 mg/kg PO q 14–21 d
Ivermectin	2 mg/kg topically, then rinse off after a few minutes; 0.2-0.4 mg/kg PO once; 0.2-0.4 mg/kg PO or IM q 14 d
Levamisole	8–10 mg/kg IC or topically q 14–21·d; 10 mg/kg IM q 14 d; 50–100 mg/L bath for 1–8 h q 7 d; 100–300 mg/L bath for 24 h q 7–4 d for a minimum of 3 tx; 100–300 mg/L bath for 72 h q 14–21 d for a minimum of 3 tx
Mebendazole	20 mg/kg PO q 14 d
Praziquantel	10 mg/L bath up to 3 h q 7–21 d; 8-24 mg/kg PO, SO, IC, or topically q 7–21 d

Compounds for Anesthesia, Analgesia, and Euthanasia

Benzocaine	50 mg/L (larvae); 200–300 mg/L (adult), buffered with sodium bicarbonate to adjust pH to around 7.0–7.4 (sedation, anesthesia)
Buprenorphine	75 mg/kg SQ (for analgesia)
Dexmedetomidine	40 mg/kg SQ (for 4+ h analgesia); 120 mg/kg SO (for >8 h analgesia)
Fentanyl	1.0 mg/kg SQ (for analgesia)
Ketamine/Diazepam	20–40 mg/kg ketamine/0.2–0.4 mg/kg diazepam IM (sedation, anesthesia for short procedures)
Meperidine	100 mg/kg SQ (for analgesia)

TABLE 8 (CONTINUED) SELECTED DRUGS AND COMPOUNDS SUGGESTED FOR USE IN CAPTIVE AMPHIBIANS

Morphine	38 mg/kg SQ once (for analgesia); 30–100 mg/kg SO, IM or topical (for analgesia)
Sodium pentobarbital	60 mg/kg IC; 60 mg/kg IV (euthanasia)
Tricaine methanesulfonate	200–500+ mg/L bath, buffered with an equal amount of sodium bicarbonate to adjust pH to 7.0–7.4
(MS-222)	(sedation, anesthesia); 10 mg/L bath for minimum of 30 min followed by cervical separation (euthanasia)

Vitamins

Vitamin B-1	25 mg/kg PO prn; 50–100 mg/kg IM or IC prn (thiamine)
Vitamin B complex	0.5–1.0 mL/gal water bath
Vitamin D$_3$	2–3 iu/mL continuous bath; 100–400 iu/kg PO q 24 h; 1000 iu/kg IM once 24 h after resolution of tetany following Ca tx
Vitamin E	1 mg/kg IM or PO q 7 day

Hormones

Gonadotropin-releasing hormone	0.1 mg/kg SC, IM, repeat prn (induction of ovulation)
Human chorionic gonadotropin	250–400 IU SC, IM (induction of ovulation); 50–100 IU SC, IM (sperm release); 300 IU SC, IM (sperm release)
Pregnant mare's serum gonadotropin	50 IU SC, IM (induction of ovulation)
Progesterone	1–5 mg SC, IM (induction of ovulation)

Miscellaneous Compounds

Allopurinol	10 mg/kg PO q 24 h (for gout)
Dexamethasone	1 mg/kg IM (for shock)
Epinephrine 1:1000	1 drop/100 g (for respiratory problems and cardiac arrest)
Prenisolone sodium	5–10 mg/kg IM (for shock) succinate

[a] IC, intracoelomic; IM, intramuscular; PO, per os; SQ, subcutaneous; tx, treatment.
Source: This compendium originally appeared in *ILAR Journal* 48(3). It has been adapted with the author's permission and is reprinted with permission from the *ILAR Journal*, Institute for Laboratory Animal Research, The National Academies, Washington, D.C. (www.nationalacademies.or/ilar).

blood smears, by microbial culture of heart blood, liver, and coelomic fluid, and by environmental testing.

- For best results, the organisms require prolonged incubation (18 to 24 hours at 30°C) on various types of agar plates (such as 5% sheep blood, chocolate agar, and MacConkey's) before the characteristic yellow, beta-hemolytic colonies appear. Biochemical testing, antimicrobial sensitivity testing, and molecular diagnostics (PCR and PFGE) may be required for further characterization and strain determination.

- Treatment is not recommended. Frogs showing clinical signs should be culled to prevent the spread of the disease.
- *Chryseobacterium* is a potential zoonotic pathogen. Gloves should be worn when handling frogs, equipment, and tank water.

Mycobacterium spp.

- *Mycobacterium* are small, acid-fast organisms ubiquitous in aquatic environments. Bacteria from this genus are a common cause of infectious disease in laboratory *Xenopus*.
- *Mycobacterium* are opportunistic pathogens associated with non-healing dermal ulcerative lesions, systemic disease, and visceral granulomas (not tubercles) in amphibians.
- Mycobacteria reported to cause disease in laboratory *Xenopus* include *M. xenopi, M. marinum, M. chelonae* (Clothier and Balls 1973; Green et al. 2000; Schwabacher 1959), *M. fortuitum, M. liflandii* (Godfrey et al. 2007; Suykerbuyk et al. 2007; Trott et al. 2004), *M. szulgai* (Chai et al. 2006), and *M. gordonae* (Sánchez-Morgado et al. 2009).
- The clinical signs include focal or multifocal non-healing ulcerative skin lesions (**Figure 25 A and B, Figure 26 A**), lethargy, coelomic distention and large granulomatous liver lesions (**Figure 26 B and C**), and weight loss. Mycobacteriosis can be a chronic, progressive disease that can present a wide range of symptoms. It may suddenly appear with fulminate clinical signs, or result in sudden death in the absence of discernable gross lesions. Infections can be sporadic, affecting a single animal (as reported for *M. chelonae*), or the disease can have a high prevalence once infection is established (*M. liflandii, M. marinum, M. szulgai* and others).
- A provisional diagnosis can be aided by special stains detecting acid-fast bacilli in tissue section (**Figure 26 D**). However, staining can be insensitive (false negatives can occur) and the definitive diagnosis may best be determined by microbial culture and molecular diagnostics. Microbial culture, however, is time-consuming and can be difficult because the organism is fastidious and very slow growing. As an alternative, PCR can provide a timely and reliable diagnosis (Densmore and Green 2007).

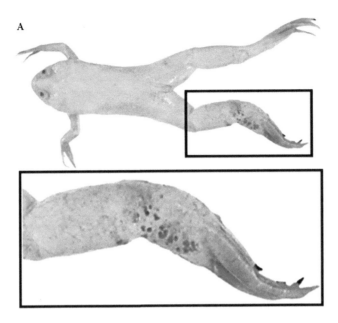

Figure 25 (A) Ulcerative skin lesions on an icteric albino laboratory *Xenopus* due to *M. chelonae* infection. (From Green, S. L. et al., *Comp. Med.* 50(6): 675–679, 2000. With permission.)

- Treatment is not recommended because of the risk of spreading the infection. Affected animals should be culled and tanks and associated equipment thoroughly cleaned and sanitized.
- Mycobacteria are potential zoonotic pathogens. Gloves should be worn when handling frogs, equipment, and tank water.

Chlamydia spp.

- *Chlamydia* are Gram-negative, obligate intracellular coccoid organisms. The organism is commonly found in wild and captive populations of anurans.
- *C. psittaci* and *C. pneumoniae* are the most frequently identified species associated with infection and disease and are associated with granulomatous and fibrinopurulent lesions (Bodetti et al. 2002; Jacobson et al. 2002; Reed et al. 2000).
- *Chlamydia psittaci* caused mass mortalities (20,000 to 40,000 frogs) in laboratory *X. laevis* fed raw beef liver in the 1980s

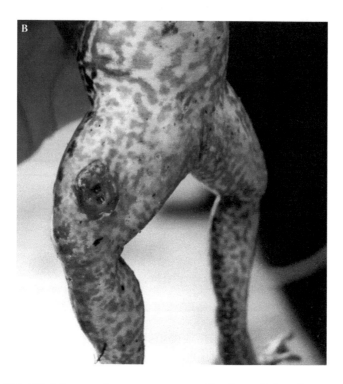

Figure 25 (B) A large ulcer, as seen on the leg of this frog, is consistent with *Mycobacterium* infection.

(Howerth 1984; Howerth and Pletcher 1986; Newcomer et al. 1982; Wilcke, Jr. et al. 1983).

- *C. suis* and *C. abortus* have been associated with subclinical infections and mortality in captive *Xenopus laevis* in Switzerland (Blumer et al. 2007).

- The clinical signs related to pneumonia and sepsis include petechiation, skin slough, hydrocoelom, and bloat due to accumulation of fluids in the subcutaneous spaces, lethargy, and the inability to dive. Red leg syndrome has also been reported in association with this infection, and *Chlamydia* are often co-cultured with numerous other opportunistic pathogens.

- At necropsy, internal organs may be grossly swollen and show histological evidence of marked histiocytic or granulomatous inflammation. Characteristic intracytoplasmic basophilic inclusions typical of *Chlamydia* are present in the hepatoyctes (**Figure 27**) (Densmore and Green 2007).

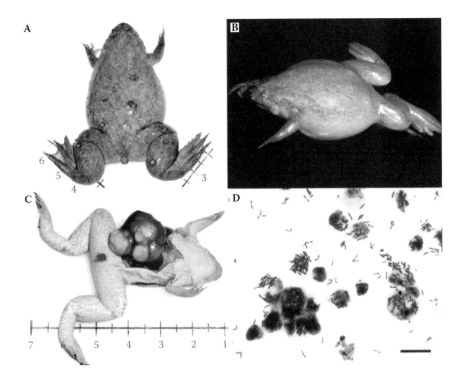

Figure 26 Two adult female *X. tropicalis*, one with severe cutaneous mycobacteriosis (A) and the other with systemic mycobacteriosis (B). In (A), note the multifocal cutaneous ulcers. Bar = 10 mm. In (B), note the coelomic effusion and generalized edema. (C) The liver from an infected *X. tropicalis*. Note the hepatic granulomas and the focal cutaneous ulceration. (D). Cytologic examination of the coelomic fluid from an *X. tropicalis* with severe coelomitis and massive effusion. Numerous paired and individual mycobacteria are seen within macrophages and free within the effusion. Ziehl-Neelson acid-fast stain; bar = 14 μm. (From Trott, K. A. et al., *Comp. Med.* 54(3): 309–17, 2004. With permission.)

- The diagnosis is based on one or more of the following: microbial culture, histological examination of tissues, immunohistochemistry, PCR, and ultrastructural examination of the liver showing the presence of reticulate and dense elementary bodies (Taylor 2001a).

- There is no treatment. Affected animals should be culled from the colony and equipment and housing tanks thoroughly sanitized.

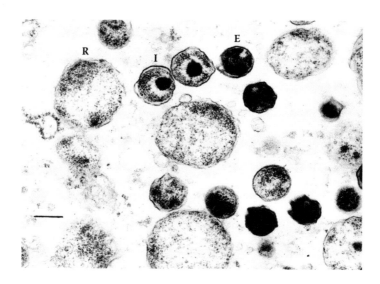

Figure 27 Transmission electron micrograph of chlamydial particles in liver from a captive African clawed frog, *X. tropicalis*. Note the
reticulate bodies (R), intermediate bodies (I), and highly condensed
elementary bodies (E). Bar = 270 nm. (Image reproduced from http://
www.cdc.gov/ncidod/eid/vol6no2/pdf/reed.pdf)

Viral Infections

Viral diseases are poorly characterized and under-recognized in
laboratory *Xenopus*. To date, there are no published reports of viral
epizootics in a population of laboratory *Xenopus*; however, as awareness increases and worldwide commerce continues to expand, viral
infections in laboratory *Xenopus* are likely to emerge. In the wild,
amphibian mortality due to infection with *Ranavirus* is quite common and is therefore described below. *Ranavirus* should be first on
a laboratory animal veterinarian's differential diagnosis list if the
history, clinical signs, and electron microscopy findings indicate a
viral disease.

Ranavirus

- Ranaviruses, family Iridoviridae, have been implicated in
  amphibian deaths worldwide, and *Xenopus laevis* has been
  identified as a potential viral reservoir (Robert et al. 2007).
  Infection with *Ranavirus*, a double-stranded DNA virus, is
  well described and has been attributed to epizootics due to
  tadpole edema virus (TEV), frog erythrocytic virus (FEV), and
  amphibian ulcerative skin syndrome.

- *Ranavirus* infection tends to cause disease in the early life stages of amphibians. The viruses are rarely isolated from clinically normal adult individuals, and infection is usually transmitted by cannibalism among larva or ingestion of affected individuals, or via contact with the free virus in the water.
- Clinical signs range from sudden death with no clinical signs to severe erythematous skin lesions, paint-brush hemorrhages, vesicular or erosive skin lesions, and related signs of sepsis (buoyancy deficits, lethargy, and bloat).
- There is no effective treatment. Affected animals should be culled from the colony to limit the spread of the infection. Tanks and equipment should be thoroughly sanitized.
- At necropsy, there may be hemorrhage or discoloration of numerous internal organs. Histologically, there may be diffuse necrosis of the liver and hematopoietic tissues. Basophilic intracytoplasmic inclusions typical of *Ranavirus* may be evident in the liver (Bollinger et al. 1999).
- The diagnosis is confirmed by cell culture, PCR, and light and electron microscopy.

Lucke herpesvirus

- Lucke herpesvirus is an oncongenic virus first described in association with a spontaneous renal carcinoma from a northern leopard frog (*Rana pipiens*) (Lucke 1934).
- At the time of this publication, Lucke's renal carcinoma occurs spontaneously only in the northern leopard frog, *Rana pipiens*.
- Renal carcinomata have been reported in captive *X. laevis* (Elkan 1960). However, the definitive identity of the virus associated with the tumors in *Xenopus* was not determined.

Fungal Infections

Batrachochytrium dendrobatidis

- Chytridiomycosis is caused by *Batrachochytrium* spp., a ubiquitous, sporozootic fungus that is keratinophilic or chitinophilic and found in aquatic environments. This agent is a global threat to amphibians and is currently the cause of severe population declines in a broad host range of wild frogs. Wild *Xenopus* appear to be silent carriers of this disease. They do not show clinical signs and do not experience the sudden die offs related to infection with this organism as reported in other amphibian populations (Weldon et al., 2004). *Xenopus laevis* may have some innate resistance to the disease, as demonstrated by in vitro experiments using peptides secreted from their skin (Rollins-Smith et al., 2009). Release or escape of captive *Xenopus laevis* into wild populations has been associated with the spread of the disease (Weldon et al., 2004).

- There are many genera and species of chytrid fungi, but *Batrachochytrium dendrobatidis* is the only chytrid known to infect vertebrates.

- Disease and significant mortality due to infection with *Batrachochytrium dendrobatidis* is well recognized in captive laboratory *Xenopus*, particularly laboratory *Xenopus tropicalis* (Mazzoni et al. 2003; Parker et al. 2002; Pessier et al. 1999). Hill and colleagues (2009) have reported *Batrachochytrium dendrobatidis* infection in a *Xenopus laevis* concurrent with *Aeromonas hydrophila* infection.

- Clinical signs of chytrid infection in *Xenopus* include epidermal hyperplasia/hyperkeratosis and skin discoloration (**Figure 28**), abnormal skin shedding, dermal inflammation, lethargy, and dehydration. Secondary bacterial or other fungal infections may be present and result in ulcers, petechiae, and ecchymosis. Mortality rates are high. Preceding death, signs can be severe or minimal. In tadpoles the infection is characterized by loss of black coloration of the mouth parts and rounding of the cutting edges of the jaw sheaths.

- A presumptive diagnosis can be made by detection of the spheroid to slightly oval fungal spores (thalli) in wet mounts of infected skin (in shed skin and in scrapings from retained layers of epidermis on infected animals).

- The wet-mount prep positive diagnosis can be confirmed by histology, immunohistochemistry, or by culture of the organism (this requires a laboratory with mycological expertise). Chytrid thalli have periodic acid-Schiff-positive walls and stain well with argyrophilic and argentophilic silver stains (**Figure 29**). PCR-based diagnostics are now widely available and specific (Kriger et al. 2006). The organism is ubiquitous, and detecting its presence in a skin swab sample by PCR confirms its presence but does not confirm disease and infection of an individual frog.

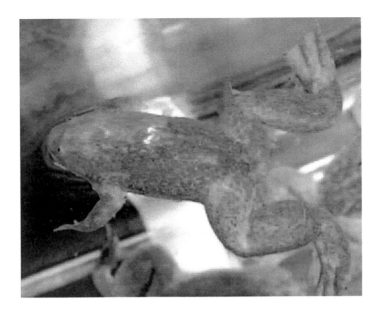

Figure 28 An adult male *Xenopus tropicalis* with severe chytridi-omycosis. Note the generalized hyperpigmentation, dysecdysis, and loss of the slime coat. (From Parker, J. M. et al., *Comp. Med.* 52(3): 265–68, 2002. With permission.)

- One report describes the treatment of laboratory frogs infected with *B. dendrobatidis* (Parker et al. 2002). However, because *B. dendrobatidis* zoospores are readily transmissible and can infect a wide range of host vertebrate species, treatment is not recommended. Infected *Xenopus* should be culled from the laboratory research colony. Equipment and surroundings associated with infected animals must be cleaned and disinfected with antifungal agents to prevent persistence of the infection and potential dissemination. Incoming shipments of frogs should be quarantined for observation prior to introduction into a laboratory colony.

Saprolegnia spp. and other water molds

- Saprolegniasis, a common infectious disease of *Xenopus*, is caused by the water mold *Saprolegnia* spp. (or similar agents such as *Achlya, Leptolegnia,* and *Epistylididae*) (Densmore and Green 2007; Pritchett and Sanders 2007).
- Water molds are frequently involved in secondary skin infections of debilitated, injured, or diseased aquatic species and are

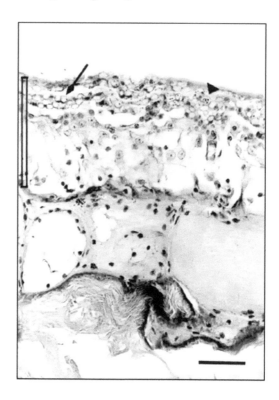

Figure 29 Photomicrograph from a skin section of an African clawed frog, *X. tropicalis*, with chytridiomycosis. Note the numerous thalli of *B. dendrobatidis* present in the cytoplasm of keratinocytes in the stratum corneum. The thin arrow points to empty thalli. Some are filled with zoospores (arrowhead). There is severe hyperkeratosis, acanthosis, spongiosis, and intracellular edema of the epidermis. The vertical bar indicates the relative thickness of the epidermis. H&E stain; bar = 50 μm. (From Parker, J. M. et al. *Comp. Med.* 52(3): 265–68, 2002. With permission.)

associated with poor water quality and water with large amounts of organic debris. *Saprolegnia* in particular are known to infect fish and can be transmitted between fish and amphibians.

- *Saprolegnia* can infect harvested *Xenopus* eggs and oocytes, diminishing the quality and utility of the material for research.

- Clinical signs of fungal infection in *Xenopus* include lethargy and grossly visible grayish-white tufts of cottony mold loosely attached to the skin of the animal (**Figure 30**). Mats of the organisms can often be seen floating in the water. Mortality is

Figure 30 Cottony tufts of the water fungus *Epistylis* attached to the toes of a laboratory *Xenopus*. (From Pritchett, K. R. and G. E. Saunders, *J. Am. Assoc. Lab. Anim. Sci.* 46(2): 86–91, 2007. With permission.)

variable, but is generally low and often depends on the underlying disease or problem. Death may occur in heavy infection (covering a large percentage of the body) presumably from osmoregulatory impairment. Inflammatory response is minimal, but some lesions are associated with ulceration, necrosis, and edema.

- The diagnosis can be made by examining fungal-infected tissues by wet mounts and detection of mats of aseptate, sparely branching fungal filaments (**Figure 31**). Histologically, fungal filaments and zoospores are evident in lesions. The diagnosis can be aided by microbial culture and molecular diagnostics (Taylor 2001b).

- Many treatments have been recommended (Taylor 2001b; Wright 2009) and include: adding benzalkonium chloride (1:4,000,000 or 0.0025 mg/L) to the tank water, then changing the water three times weekly; copper sulfate (1:2000 or 500 mg/L) immersion for 2 minutes daily for 5 days until healed; immersion in potassium permanganate (1:5000 or 200 mg/L) for five minutes (Taylor 2001b); once weekly 30-minute baths in noniodized sodium chloride (25g/L); malachite green baths (67g/L) for no more than 15 seconds per day for 2 or 3 days. Additional antifungal agents and doses are listed in **Table 8**. However, if the underlying disease is suspected to be related to bacterial infection, the prognosis is poor.

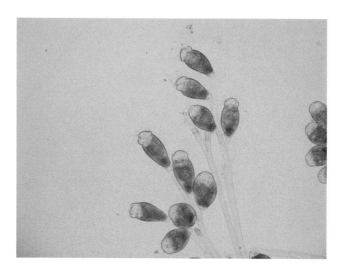

Figure 31 Photomicrograph (magnification 40X) wet mount preparation from infected frogs showing the *Epistylis* organism. Note the long
stalks with the round/cylindrical zooids. The zooids are ciliated (not
seen at this magnification). (From Pritchett, K. R. and G. E. Saunders,
J. Am. Assoc. Lab. Anim. Sci. 46(2): 86–91, 2007. With permission.)

Parasitic Infections

Xenopus may be infested with a variety of cestodes, trematodes,
nematodes, and protozoa. Indeed, more than 25 genera of parasites
have been reported from wild *Xenopus* (Tinsley 1996). Parasitism of
the captive-reared laboratory frog is almost always related to stress,
water temperature or water quality problems, or husbandry management practices (including the introduction of wild-caught frogs into
the colony and purchasing and co-housing frogs from different commercial vendors). Because wild-caught *Xenopus* may bear a significant parasite load asymptomatically, they should not be introduced
into a laboratory colony prior to antihelmintic therapy and until
after quarantine. The most common parasites reported in laboratory
Xenopus are described below.

Pseudocapillaroides xenopi (Capillaria xenopodis)

- Infection with the helminth *Pseudocapillaroides xenopi* (syn =
  Capillaria xenopodis) can cause wasting disease and significant morbidity and mortality in laboratory *Xenopus* (Brayton
  1992; Cunningham et al. 1996; Ruble et al. 1995; Stephens
  et al. 1987).

- Infection with *Capillaria* can result in skin infections that progress to septicemia. The parasite inhabits the epidermis, principally on the dorsum of the body.

- The life cycle is direct, and heavy infections may develop in 6 to 18 months in frogs that share an infected enclosure.

- Heavily infected frogs may have respiratory difficulty, may become debilitated and anorexic, and may lose weight, while the normally smooth skin on the animal's back becomes roughened, grayish, and pitted (Cunningham et al. 1996; Stephens et al. 1987). There is marked epidermal hyperplasia histologically, with inflammatory, degenerative, and proliferative changes. Ulcers may be present, and patches of skin may slough.

- With the aid of a dissection microscope, the parasite, a 2–4 mm white nematode, and tunnels from its migration can often be seen just beneath the mucous layer of the skin (**Figure 32**). Adult worms, larvae, and thin-walled, larvated bioperculate ova may be seen in the wet mount preps of flakes of skin or skin scrapings (**Figure 33**) or in water sediments (Cunningham et al. 1996; Stephens et al. 1987).

- Because the life cycle is direct, the skin flakes and parasites in the water readily infect other frogs. In an intensive housing or high-density stocking situation, the best practice is to cull and euthanize heavily infected animals. Treatment with Ivermectin® (Merck & Co., Inc., Whitehouse Station, NJ) (2 mg/kg topically for a few minutes, then rinse off) or 0.2–0.4 mg/kg PO or IM q 14 d, or other antihelmintics (**Table 8**) and various other treatments have been recommended, including oral or bath treatments with fenbendazole, levamisole, or thiabendazole (Hadfield and Whitaker 2005; Iglauer et al. 1997; Pessier 2002; Wright 2006). With treatment, there is risk of anaphylactic shock and death due to the immune responses to the massive die-off of nematodes in heavily infected animals. There have been anecdotal reports of deaths in laboratory *Xenopus* related to thiabendazole treatment.

- There are anecdotal reports of increased rates of infection related to lowering of the water temperature (from 20°–22° to 15°C) for several days in captive *Xenopus* colonies. Frogs that are stressed or chilled may have diminished immune response and be more susceptible to the parasite.

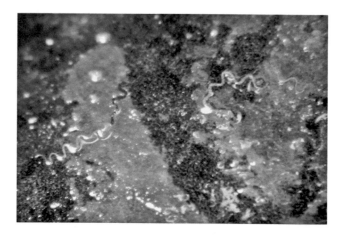

Figure 32 With the aid of a dissection microscope, the parasite *Pseudocapillaroides xenopi*, a 2–4 mm white nematode, can be seen in the skin on the dorsum of an infected laboratory *Xenopus*. Note the excessive mucous accumulation.

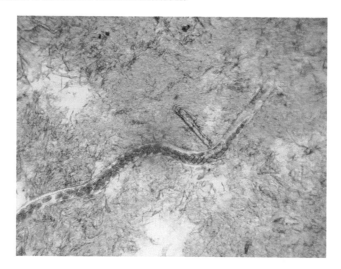

Figure 33 An adult *Pseudocapillaroides* containing eggs in a skin scrape wet mount preparation. (Photo courtesy of Mike Wages.) Magnification 3.2X.

Rhabdias (strongyloid lungworms)

- *Rhabdias* have been reported to infect laboratory anurans (Poynton and Whitaker 2001).
- Lungworms may cause pneumonia, anemia, and failure to thrive.

- The life cycle is direct. Hermaphroditic adults live in the lung, where they deposit eggs containing the first-stage larvae. These are coughed up, swallowed, and passed out in the feces. Eggs and larvae may be present in the intestines in large numbers. Larvae that have hatched in the water and penetrated the skin have been found floating freely in coelomic cavity and lymph spaces in laboratory *Xenopus* (personal observation, S. Green).
- Therapy of choice for *Rhabdias* infection in most species is Ivermectin at a dose of 0.2 to 0.4 mg/kg administered orally or subcutaneously. For *Xenopus*, the administration of the compound subcutaneously is the preferred route.
- Tank disinfection combined with repeated treatments may be necessary to achieve consistent negative fecal samples.

Cryptosporidia

- Cryptosporidiosis is rare. It has been associated with emaciation and proliferative enteritis in a two-year-old female laboratory *Xenopus* (Green et al. 2003).
- Numerous cryptosporidial stages were present throughout the intestinal tract. Oocysts were present in the aquaria water. No other frogs in the tank were affected.
- The original source of the organism was not determined, but a contaminated water source was suspected, as this organism is typically associated with water-borne illnesses in other species, including humans.
- *Cryptosporidia* is a potential zoonotic pathogen. Gloves should be worn when handling *Xenopus.* Aquaria water filters with pores of 0.01 microns or less should prevent oocysts from entering the housing system (Green et al. 2003).

Acariasis (mites)

- An unspeciated cutaneous mite (similar to those of the genus *Rhizoglyphus*) has been reported to cause mortality in adult female laboratory *Xenopus* (Ford et al. 2004). The mite is not considered to be a parasite of *X. laevis*, but instead feeds off moss, fungi, and detritus. Subsequent examination of the moistened sphagnum moss used for shipping the frogs from the vendor revealed the same mite. The authors suspect the moss was the source of the mite infestation. Mites were attached to the growth of *Saprolegnia* on the frogs.

- Frogs presented dermal erythema and petechiation consistent with red leg, and dermal ulcers with white filamentous growth on the skin consistent with *Saprolegnia*. Mite infestation was diagnosed by skin scraping and identification of a mite with piercing mouthparts and eight legs composed of four segments, each ending with a sickle-shaped claw.

- A nasal cavity mite, *Xenopacarus africanus,* is found in the nasal cavity and Eustachian tubes of *X. laevis* (Tinsley 1996). *Xenopacarus africanus* is of the family Erynetidae and in the same family as the chigger. Infestation of laboratory *Xenopus* with *Xenopacarus africanis* (which is considered commensal) has not been associated with morbidity or mortality.

- *Xenopacarus africanis* feeds off of blood from vessels in the head of the frog and may occasionally be observed exiting from the nares of laboratory *Xenopus* that have been anesthetized by immersion in MS-222 (S. L. Green, personal observation).

noninfectious diseases and conditions

Dehydration and Desiccation

- *Xenopus* are susceptible to dehydration and death by desiccation.

- Wild *Xenopus* can survive desiccation during aestivation (by burrowing into the mud and forming a slime coat protective cocoon around themselves). They might also migrate for miles over dry desert (sustained by intermittent rains) to get to a new water source.

- In contrast, laboratory *Xenopus* should not be kept out of the water for more than 15 to 20 minutes. Those that are out of the water for several hours (those that escape their housing tanks) are not likely to survive.

- The skin of a dehydrated animal turns dark, wrinkles, and is tacky to the touch. The slime coat turns opaque and grayish. Desiccated animals have leathery, dry skin.

- Dehydrated laboratory *Xenopus* are often in shock. Drugs used for treatment of shock in amphibians are listed in **Table 8**.

Dehydrated *Xenopus* should be returned to a shallow bath of clean, cool, and well-oxygenated tank-water, and then placed in amphibian Ringer's solution (**Table 7**) until recovered. Amphibian Ringer's solution can be sterilized and given IC or SQ and should be administered at a fluid volume of 2–5% of the animal's body weight. Artificial slime (Shield-X®) or water conditioners may be added to the water to soothe the dehydrated epidermis.

- To prevent dehydration during surgical procedures or the post-operative recovery period, frogs should be partially submerged (with the nose and mouth elevated and out of the water) and covered with moistened drapes.

Gas Bubble Disease

Gas bubble disease is a noninfectious syndrome seen in aquatic frogs when water is supersaturated with gases, particularly nitrogen and argon (Colt et al. 1984). In a laboratory environment, supersaturation of the water with nitrogen and argon most often occurs when air leaks into the water pipes and the water is pumped under pressure to supply the aquaria. If the water is pressurized, its capacity to hold soluble gases increases. Water drawn from a deep well and naturally supersaturated with nitrogen and argon (deep well water is under high hydrostatic pressures) was first reported as a cause of gas bubble disease in laboratory *Xenopus* in 1984 (Colt et al. 1984). Heating the water further increases its capacity to hold soluble gases (Colt 1986).

- Clinical signs of gas bubble disease in frogs include buoyancy problems, micro and macroscopic bubbles in skin and foot webbing (**Figure 34**), hyperemia in webbing and legs, petechial and ecchymotic hemorrhages progressing to skin erosions, and loss of mucous coat. An increase in the occurrence of red leg due to secondary infections with A. *hydrophila* was reported in gas bubble disease (Colt et al. 1984).

- Morbidity associated with gas bubble disease in laboratory *Xenopus* is high but many animals can recover when the water problem is corrected. Death is usually due to secondary septicemia or fatal thromboembolic events. Animals that recover may not produce good quality or significant quantities of eggs or oocytes for months thereafter.

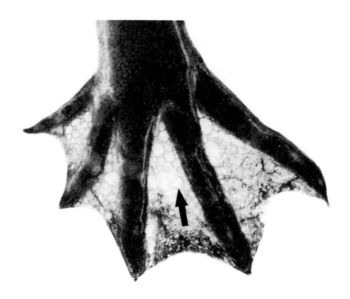

Figure 34 Macro- and microscopic gas bubbles in the skin in the inter-digital webbing of a laboratory *Xenopus* with gas bubble disease. (From Colt, J. et al., *J. Herpetol.* 18(2): 131–37, 1984. With permission.)

- **Supersaturation** is determined by measuring the sum of all the gasses dissolved in the water: the **total gas pressure** of the waste (Colt 1983).

- Under normal conditions total gas pressure of water is 100%. An increase in total gas pressure by as little as 1–2% (from 100% to 102%) will cause gas bubble disease (Colt et al. 1986).

- A tank with supersaturated water can *sometimes, but not always* be detected by observing bubbles in the water, by observing bubbles and a change in color in the water coming from the pipes (entrained air may turn the water milky white), and by observing bubbles sticking to the sides of the tank and other items in the water (**Figure 35**).

- It must be recognized that gas bubble disease is usually related to supersaturation of water with nitrogen gases (and others) rather than with oxygen. Nitrogen causes frog and fish health problems at *any saturation level greater than 100%*. Nitrogen is a primary component of the gases in the earth's atmosphere. However, unlike oxygen, nitrogen is slow to escape from the water's surface. A reduction of nitrogen gas saturation levels from 110% to 100% can take up to 5 or 6 hours or more. In contrast, oxygen saturation levels can be safe exceeding

Figure 35 Note the milky-white water entering a pond-style *Xenopus* housing tank. This water contains tiny bubbles and is supersaturated with gasses. Water with entrained air is the cause of gas bubble disease in *Xenopus*. Water that is supersaturated with gasses can also appear perfectly normal and contain no bubbles.

200%. Oxygen can readily diffuse from the water's surface and is consumed by the frogs.

- Aquaculture operations use a saturometer to monitor total dissolved gases in the water (Weiss Saturometer; from ECO Enterprises). Saturometers are expensive, take time to set up and equilibrate, and generally require an experienced technician to operate.

- In research animal facilities, dissolved oxygen (DO) measurements can be used as a general indicator of gas supersaturation. Abrupt increases in DO by as little as 2–4 mm Hg, or 2–4%, may mean supersaturation of the water not only with oxygen, but potentially with other gases as well.

- To correct supersaturation the water must be **degassed**, a process that creates a large interface between the water and the air. Degassing water is achieved by **aeration** (mixing air and water) in a variety of ways: the use of air stones, water diffusers, packed columns, weirs, pond fountains, and water-fall-type water cascades. Aeration seems counterintuitive, but the surface area of the bubbles created provides a large air-water interface where the gases can escape directly to the atmosphere. Degassing systems are not currently a feature of most modular housing systems and are best suited for pond style housing.

- Gradually heating the water to room temperature will also improve the rate at which gases move from the water to the air. This can be accomplished by increasing the environmental temperature in the frog housing room.

- Frogs showing signs of gas bubble disease should be treated by isolation into aquaria tanks filled with degassed, normal water. Water conditioners should be added daily (Shield-X®, Novaqua®*, Amquel®) to the water to aid in the prevention of secondary skin infections and preserve the slime coat layer. Animals with petechia and hemorrhage and bubbles in the skin may recover in a week or more; however, females may not produce quality eggs or oocytes for many months.

- Those in charge of *Xenopus* aquaria should keep in contact with municipal water suppliers and facility/operations managers, and subscribe to receive weekly municipal or campus water quality reports. Request notices of changes to the mainline water system (repairs, shutdowns, diversions, changes in water sources) or other manipulations that may introduce air into the waterlines.

Chlorine/Chloramine Toxicities

- *Xenopus* skin is permeable and is particularly sensitive to many superficial irritants, including ammonia, pesticides, disinfectants and detergents, heavy metals, chlorine and chloramines.

- Exposure to high levels of these agents can cause lethargy, heavy skin sloughing (**Figure 36**), and death.

---

* (Kordon LLC, Division of Novalet TIC, Hayward, CA)

Figure 36 Skin sloughing in laboratory *Xenopus* as a result of exposure to high levels of chlorine in the water. Arrows are pointing to shed skin.

- Chloramine and chlorine toxicity can be associated with eye irritation, gill irritation in tadpoles, skin sloughing, and acute death *in the absence* of other symptoms.

- Chlorine and chloramines are oxidizing agents. At high levels of concentration, red blood cells are destroyed, and the iron present in hemoglobin becomes oxidized so that the oxygen-carrying capacity is diminished (Whitaker 2001).

- Dechlorinators containing sodium thiosulfate will rapidly remove chlorine. Aeration of the water will also remove the chlorine. However, chloramine is more toxic and more stable, and conditioners designed to remove it often cause the release of more ammonia.

- Activated carbon alone will not remove the chloramines. Chloramine removal is best accomplished using carbon filters charged with an ammonia-absorbing material such as the mineral clinoptilolite (Whitaker 2001).

- Chlorine/chloramine levels should be immediately checked whenever there are acute deaths in the colony with no obvious symptoms or if large amounts of shed skin are in the water. If levels of chlorine/chloramines are high, the water should be changed immediately or the animals placed in amphibian Ringer's solution (**Table 7**), and water conditioners or artificial slime (Shield-X®) should be added to chlorine/chloramine-free replacement water to soothe the injured frogs' epidermis.

Neoplasia

- Spontaneous neoplasia occurs sporadically in *Xenopus*. The disease is infrequently reported, but lymphosarcoma and melanophoroma are most often described (Green and Harshbarger 2001). **Table 9** summarizes the types of spontaneous neoplasia reported in *Xenopus*.

Rectal and Cloacal Prolapses

- Cloacal and rectal prolapses may occasionally be observed in laboratory *Xenopus* and may be secondary to gastric overload, neoplasia, parasites, or ascites and coelomic accumulation of

TABLE 9 SPONTANEOUS NEOPLASMS REPORTED IN CAPTIVE XENOPUS

Neoplasm	Site of Origin	N=
Lymphosarcoma	Liver, spleen, mesonephros, viscera, thymus	13
Melanophoroma	Skin on the leg, back, snout	9
Epithelioma	Head	>1
Fibroma	Coelom, head	2
Lipoma	Fat bodies, urinary bladder	2
Granular Cell Tumor	Head	1
Nephroblastoma	Mesonephros	1
Renal Cell Carcinoma	Mesonephros	1
Hepatoblastoma	Liver	1
Bile Duct Adenoma	Liver	1
Adenocarcinoma	Intestine	1
Adenoma, Papillary	Harderian gland	1

Source: Adapted from Kevin M. Wright and Brent R. Whitaker, *Amphibian Medicine and Captive Husbandry* (Malabar, FL: Krieger Publishing Company, 2001), pages 385–96.

fluids; or, in the case of females, cloacal prolapse can occur after "priming" with PMSG or be due to trauma associated with an overly rough "milking" or squeezing of the animal as a means of collecting eggs and oocytes.

- Reduction of the prolapse is achieved by applying a hyperosmotic saline (2–5%) ophthalmic solution, or a hyperosmotic sugar solution. The tissue swelling should decrease after a few minutes. It can then be coated with a water-soluble lubricant and replaced back through the cloacal opening using cotton tip applicators.

Gout

- Gout is due to renal dysfunction and the accumulation of urate crystals in the joints, usually in the digits. A painful, chronic inflammatory response ensues, resulting in tissue swelling (usually at the tip of a digit).
- Gout in anurans may be related to high-protein diets, prolonged periods of dehydration, ingesting prey that fed upon plants high in oxalates, or exposure to aminoglycosides.
- Idiopathic gout has been observed by the author in a 4-year-old female laboratory *Xenopus*. The affected digit was successfully amputated and the frog lived another year without recurrence.

Skeletal Deformities

- Scoliosis due to nutritional or possible genetic defects and multi-limbs or "spindly limbs" have been infrequently observed by the author in large shipments of vendor-supplied, captive-reared laboratory *Xenopus*.
- Skeletal abnormalities, anemia, and weight loss have been associated with hypervitaminosis A in *Xenopus* fed mammalian liver or whole immature rodents, both of which are high in vitamin A (Densmore and Green 2007).
- Rickets and osteoporosis have been associated with vitamin D deficiency in *Xenopus laevis* exclusively fed horse livers (Bruce and Parkes 1950).
- Suggested doses for vitamin supplementation can be found in **Table 8**.

Bite Wounds

- Bite wounds can be observed on the legs and axilla of frogs housed in densely stocked tanks, particularly in the absence of refuge cover and enrichment.

- Erosive lesions in the axilla of the front legs or at the joints of the legs (**Figure 37**) can occur as a result of one frog swallowing the limb of another during acts of aggression or during a feeding frenzy.

- The lesion can become secondarily infected with opportunistic bacteria or saprolegnia fungus.

- If severe enough, the erosive lesions extend into the muscle layer and result in loss of use of the limb and debilitation. Severely affected frogs should be euthanized.

- Frogs with minor bite wounds should be isolated in amphibian Ringer's solution (**Table 7**) for 7 to 10 days to allow the wound to heal and the animal to have unfettered access to nutrition. Water conditioners or artificial slime added to the water may speed recovery.

Figure 37 A severe bite wound in the axilla of an emaciated laboratory *Xenopus*. Laboratory *Xenopus* may swallow the arm of another frog during the feeding frenzy or as an act of aggression. (From Torreilles, S. L. and S. L. Green, *J. Am. Assoc. Lab. Anim. Sci.* 46:33–37, 2007. With permission.)

- When bite wounds are a problem, the frequency and feeding amount should be increased, the stocking density in the tank decreased, and refuge cover in the form of pipes or other containers added to the tank (Torreilles and Green 2007).

Ovarian Hyperstimulation Syndrome

- Ovarian hyperstimulation syndrome (OHS) is the result of a rare response to hormonal priming of laboratory *Xenopus* for the purpose of inducing ovulation and egg laying (Green et al. 2007).
- Affected female frogs typically become greatly bloated (**Figure 38**) due to accumulation of the fluids beneath the skin in the large subcutaneous space, 3 to 14 days after administration of the hormones.

Figure 38 A bloated female laboratory *Xenopus laevis* with fluid beneath the skin due to ovarian hyperstimulation syndrome. This symptom, often referred to as hydrops, can be caused by a number of other problems, such as renal disease, sepsis, and cardiovascular and lymphatic disease. (From Green, S. L. et al., *J. Am. Assoc. Lab. Anim. Sci.* 46(3): 64–67, 2007. With permission.)

- Treatment can be attempted by placing the frog in a mildly hypertonic saline solution (0.7–0.85% saline), but the prognosis is poor. At post-mortem, a large number of eggs are often found distributed throughout the coelomic and thoracic cavity or adhered to the liver capsule or pericardial sac.

- The pathophysiology of OHS is thought to be related to hormonally induced disturbances of vasoactive mediators, including the vascular endothelial growth factor secreted by theca and granulosa cells.

Thermal Shock

- Death due to thermal shock has been reported in cold-acclimated *Xenopus* that were inadvertently returned to warm-water housing (Green et al. 2003).

- *Xenopus* can tolerate gradual temperature changes across a wide range of temperatures; however, sudden increases or decreases of as little as 2° to 5°C can result in thermal shock and death.

- Thermal shock usually results in acute death of the warm- or cold-acclimated frog (within 5 or 10 minutes of exposure) to cold or warm water, respectively.

Poor Egg Production, Poor Egg Quality

Under ideal conditions, egg production in laboratory *Xenopus* can occur year round. However, laboratory *Xenopus* may go through periods (months) of unexplained inefficient or complete failure of egg production, or the animals may produce eggs of inferior quality. This results in experimental delays, inability to reproduce experimental results, and ultimately, the purchase and use of more animals. Adult female laboratory *Xenopus* that are 2 to 3 years of age should produce several hundred to several thousand eggs per clutch. Occasional degenerate eggs are not uncommon, but degenerate eggs should make up a very small percentage (< 2%) of the total eggs in the clutch. Degenerate eggs are small and sunken, wrinkled, shriveled, and irregularly pigmented and cannot be used for biochemical experiments or microinjection. As has been reviewed extensively (Green 2002), once infectious and parasitic diseases in the female have been ruled out as a reason for poor quality eggs and poor production, the following factors affecting oogenesis should be investigated further.

Temperature—Gonadal development and oogenesis are under the control of the adenohypophysis (pars distalis of the pituitary gland) (McLaren 1965). Lower water temperature has a negative effect on secretion of gonadotropins by the pituitary gland, on the sensitivity of the germinal epithelium, and on vitellogenesis and the synthesis of yolk. In the wild, water temperature declines in the fall and winter (a time when many researchers report declines in egg quantity and egg quality). Most laboratory-reared, egg-producing female *Xenopus laevis* are kept year round in water temperatures ranging from 19° to 24°C. There is also evidence that vitellogenesis and oocyte growth are regulated by estrogen, and estrogen increases in cold-adapted frogs as they divert energy and nutrients to oocyte production during hibernation in preparation for the spring breeding season (Duellman and Trueb 1994). *Xenopus* may actually benefit from periods of cold exposure to promote the production of estrogen and metabolic conditions necessary for oogenesis and ovulation. This may explain why some researchers report improvement in egg quality and quantity after frogs are kept in water at 16° to 17°C and gradually reintroduced into warmer temperature prior to collecting eggs. Frogs are stressed, however, at being kept at temperatures below 14°C. Water temperatures above 30°C can be lethal to *Xenopus* adults, eggs, and embryos.

Age of the female—By 2 to 3 years of age (snout-to-vent length ~80–140 mm) laboratory *Xenopus* have reached their peak reproductive period and can lay 3 to 4 clutches per year. The fecundity of laboratory *Xenopus* around 4 to 5 years old and older does not appear to be very good (L. Northerby, NASCO Inc., personal communication). Most egg-laying laboratory *Xenopus* are retired by 5 to 6 years of age, if not before.

Nutrition—Ovary production in *Xenopus laevis* is affected by food supply: during food shortages, the ovaries regress. *Xenopus* are remarkably tolerant of starvation and can reduce energy consumption for many months during hibernation or adverse conditions and will use the gonads as an additional energy reserve. Use of this energy reserve may occur in the absence of overt weight loss. While no feeding practice is proven to optimize fecundity of laboratory *Xenopus*, diminished food supply due to inadequate amounts, poor quality feed, infrequent feedings, or increased competition in the tank among individuals will diminish the quality and quantity of eggs produced. As a general rule, frogs should consume all feed within two hours after feeding, and if food is rapidly eaten (within 5 to 20 minutes or less) during the feeding frenzy, more should be fed.

Egg-harvesting methods—Overharvesting of eggs or oocytes (collections more frequently than every 3 or 4 months) may result in a decrease of good quality eggs and overall egg quantity. Alternately, maintaining frogs for long periods without harvesting eggs will result in an increase in the percentage of degenerating follicles in the ovary and eggs that develop poorly. Leaving eggs to sit too long before use or harvesting using unclean surgical techniques can result in degeneration due to contamination of eggs with bacteria (Elsner et al. 2000) or *Saprolegnia* (Green 2001).

Poor water quality—Exposing *Xenopus* eggs, embryos, and larvae to ammonium nitrate, sodium nitrate, or urea results in degeneration and mortality (Schuytema and Nebeker 1999).

diagnosis of infectious diseases in laboratory *Xenopus*

- In the face of a disease outbreak, sick and moribund frogs should be culled from the colony immediately. Tank water samples should be collected and water quality tests should be performed.

- Fresh postmortem examinations should be obtained on 5 or 6 frogs that have the typical clinical signs.

- Heart-blood and spleen should be aseptically collected and cultured for opportunistic bacterial pathogens.

- Whole blood smears should be prepared and examined for the presence of bacteria, parasites, and toxic changes in the neutrophils.

- Focal and nodular lesions in the skin, muscle, or viscera especially should be Gram's and acid-fast stained and cultured.

- Formalin fixed tissues (liver, blood, ulcerated skin lesions, nodular lesions, etc.) should be acid-fast stained and examined for mycobacteria.

- Save formalin-fixed tissues for future examination (e.g., electron microscopy).

- Rapid-freeze and store fresh tissue samples for future culture and DNA extraction.

- A fecal exam can detect protozoan and metazoan parasites, but significance depends upon the type and extent of infestation with presence of clinical signs.
- Skin scrapings are useful for detecting fungal, bacterial, and parasitic infections.
- Radiology can be useful for finding skeletal deformities, gastrointestinal impactions, foreign bodies, and pneumonia.
- Identification of the causative agent will most likely require a combination of histopathology, microbiological culturing, and molecular methods (HPLC, PFGE, PCR).
- Work with infectious disease specialists, microbiologists, and pathologists who are familiar with aquatic pathogens.

general comments on the treatment of infectious diseases

Treatment of laboratory *Xenopus* showing signs of infectious disease is controversial. There are numerous reports describing adding antibiotics to the tank water or administering antibiotics systemically (via PO, whole body immersion, or IM IC, or SQ injections) to frogs suspected of having a local or systemic bacterial infection. However, drug doses recommended to date for use in *Xenopus* have been based on mammalian or terrestrial or semi-aquatic amphibian pharmacokinetic studies and are not broadly applicable to fully aquatic amphibian species. In addition, little is known about the minimal inhibitory antibiotic concentrations effective against common *Xenopus* pathogens. The possibility of toxicity or of inducing microbial resistance should not be overlooked. The author has observed microbial resistance of the waterborne opportunistic pathogen, mycobacterium, when antibiotics were periodically added to the water in an attempt to control a disease outbreak.

The unique characteristic of *Xenopus* to regurgitate their stomach and remove the stomach contents in response to stress, or if the drugs or compounds administered *per os* are noxious, poses an additional therapeutic challenge. Gastric absorption of most agents is generally not reliable in *Xenopus*. *Whole body* immersion in therapeutic "bath" solutions is a convenient way to treat large numbers of animals but can cause skin irritation and sloughing of the slime coat, depending on the agent and the length of exposure time. Regardless of the route of administration, therapeutic substances will be excreted by *Xenopus* back into the water, and some of the excreted drug may still be in the form of active metabolites, many of which will be resorbed or have adverse effects on the biological filter. Furthermore, the effects of antibiotics on many experimental results are unknown. The risk of skin irritation and renal toxicity should be considered.

Lastly, in a laboratory environment, where hundreds or thousands of frogs may be affected during a disease outbreak, it is often impractical to treat, and the presence of so many sick frogs in multiple aquatic rooms risks spreading the disease. Many of the treatment regimens recommended in amphibian textbooks are intended for the treatment of a small number of animals in a zoo or exhibit environment, where individuals are rare specimens, which can be completely isolated, and which are accustomed to frequent handling. Laboratory female *Xenopus* that have survived an acute or chronic infectious illness (with or without treatment) may produce poor-quality and fewer eggs for many months thereafter, thus limiting their utility for research (personal observations, S. Green).

Antibiotic treatment, if undertaken, should be based on identification of the pathogen via molecular methods, microbial culture, and sensitivity testing on samples collected at post-mortem examination from freshly euthanized affected animals. **Table 8** includes empirical doses of selected antibiotics for the rare instances in which treatment might be warranted (MICs and antibiotic sensitivity patterns have been determined; the animals to be treated are few in number, rare specimens (transgenic founders), and can be isolated from the rest of the colony.

Additional palliative therapy for ill amphibians should include the following:

1. Move the sick frog to single cage housing so it does not have to compete for food.

2. Treat sick frogs with Shield-X® or Polyaqua® (Kordon LLC, Division of Novalet TIC, Hayward, CA) to aid damaged epithelia in preserving water and electrolytes. Add water conditioners (Amquel® or Novaqua®) to the housing water.

3. House sick frogs in amphibian Ringer's solution (**Table 7**). Change the solution daily. Aquatic-amphibian facilities should keep freshly prepared stock solutions of amphibian Ringer's solution on hand.

4. Amphibian Ringer's solution is the treatment of choice for most conditions requiring fluid therapy. Lactated Ringer's (273 mOsm/kg), Plasmalyte-A (294 mOsm/kg; Baxter Pharmaceuticals) are hypertonic for *Xenopus* (plasma osmolality 200–233 mOsm/kg), are not balanced for restoring amphibian extracellular or intracellular fluid compartments, and are *not* the treatment fluids of choice for dehydrated, ill frogs. Treatment with these solutions may be attempted (and are indicated) if evidence of volume overload is present: coelomic distention, bloat, or SQ fluid accumulation. Constant monitoring of frogs placed in hypertonic immersion baths is advised. Immersion in hypertonic saline, 0.85% (8.5 g NaCl/L water) is an alternative fluid if the others are not available.

5. In general, frogs recovering from mild red leg disease, gas bubble disease, mild dehydration, or physical injury (bite wounds) are most likely to recover; however, most will require 10 days to 2 weeks before they are well enough to be returned to regular housing.

To curtail outbreaks and the spread of infectious disease in a large *Xenopus* colony, sick frogs should be culled immediately and euthanized. Decreasing the stocking density by eliminating sick frogs or frogs that are no longer needed is also helpful. Strict attention should be paid to sanitation practices and quarantine procedures. Tanks housing affected animals should be depopulated if possible and cleaned. Equipment (nets, buckets, etc.) should be sanitized or thrown away and replaced.

Full necropsies are warranted on all affected animals so that the causative agent can be identified and tracked and husbandry practices altered to minimize the pathogen's spread. An emergency disease-response plan is highly recommended for aquatic facilities. In some cases, complete depopulation, sanitization, and repopulation with new animals from a healthy source vendor population is the most humane and economical option.

treatment of general trauma and abrasions

Laboratory *Xenopus* may sustain general trauma and abrasions to the legs, feet, or snout. The abrasion may extend from the skin into the underlying dermis and muscle (particularly if the wounds are bite wounds). If the wound is not severe or debilitating (if the animal can still swim), there is a fair prognosis for recovery. Palliative treatment should include those steps listed above: isolation into a separate tank (provide refuge coverage) and housing the animal in amphibian Ringer's solution for 10 days to 2 weeks. Water conditioners or Shield-X® are recommended to improve healing time. The animal should be fed and the water changed according to the regular schedule. The animal should be checked daily. The skin may heal and leave a white scar; however, it is generally at that time safe to return the animal to regular housing.

anesthesia

Xenopus have all of the neuroanatomical pain pathways as seen in mammalian species, and thus, like mammals, they are capable of experiencing pain (Machin 1999; Machin 2001; Stevens et al. 2001). Anesthesia is therefore required for prolonged or invasive procedures. Ice baths and rapidly chilling the adult frog are painful procedures for the animal and thus are *not* acceptable means of anesthesia. Because *Xenopus* can exchange CO_2 cutaneously and can hold their breath for prolonged periods, and because the trachea is very short and easy to perforate, intubation and the use of inhaled gas for anesthesia or the use of carbon dioxide gas (bubbled into the tank water or filled into dry chambers) is usually an inefficient and unsatisfactory method of anesthesia for *Xenopus*.

In almost all instances, the anesthetic of choice for amphibian surgery is an isomer of benzocaine: tricaine methanesulfonate (MS-222; Argent Chemical Laboratories, Redmond, WA).

Tricaine Methanesulfonate (MS-222)

- Tricaine methanesulfonate (MS-222) is the most useful anesthetic for amphibians and can be administered via immersion in a buffered solution or by intracoelomic injection.

- MS-222 is a powder that must be dissolved in water. Solutions for bath immersion are acidic and irritating and must be buffered with sodium bicarbonate to a normal pH (7.4) before use.

- MS-222 bath immersion concentrations range from 0.1–0.5% (1–5 g/L) for adults to slightly less for tadpoles (0.1–0.2 g/L) (Gentz 2007).

- Once the animal has been immersed in the bath, sedation should occur between 10 and 20 minutes. The solution is rinsed away with fresh tank water when surgical anesthesia has been achieved.

- Anesthesia after MS-222 immersion or IC injection is judged by loss of righting reflexes, loss of the corneal reflex, and loss of the withdrawal response to toe pinch. As the anesthetic level deepens, abdominal respiration is lost, followed by slowing of gular (throat) movements, which stop completely as a surgical level is reached.

- If the animal begins to recover before the surgical procedure is complete, the animal should be returned to a 50% dilution of the MS-222 solution before proceeding, if possible (Downes 1995).

- MS-222 poses some health hazards to the anesthesiologist: reversible retinopathy has been reported in aquaculture workers who regularly came into contact with MS-222 solutions by submersing their bare arms and hands into MS-222 solutions while anesthetizing fish (Bernstein et al. 1997). Gloves should be worn when handling MS-222 and the solution prepared under a biosafety hood.

- MS-222 solutions should be protected from light and stored in the refrigerator to prevent rapid degradation. Discolored solutions should be discarded.

Alternatives to MS-222 include benzocaine gel, ketamine, and clove oil, although these drugs should be used only for short, minor procedures in *Xenopus* or to provide mild chemical restraint. In contrast, the injectable anesthetic drug Propofol® does not appear to be safe in *Xenopus* (Guenette et al., 2007; see below).

Benzocaine Gel

- The successful use of benzocaine gel (Orajel® Regular*, 7.5% benzocaine, or Orajel® Maximum Strength, 20% benzocaine) to anesthetize or euthanize amphibians has been reported

---

* (Church & Dwight Co., Inc., Princeton, NJ)

(Brown et al. 2004; Chen and Combs 1999; Kaiser and Green 2001).

- 20% benzocaine gel (Anbesol®: Wyeth, Madison, NJ) applied to the ventrum of laboratory *Xenopus laevis* has also been described for use as an anesthetic agent (0.5–1 cm length application) (Torreilles et al. 2009).

Ketamine

- Intramuscular ketamine (70–100 mg/kg) can be used for minor procedures such as radiography, but recovery can be prolonged, from 12 to 18 hours (Gentz 2007).
- Animals anesthetized with these drugs remain sensitive to pain, even at high doses. The surgical use of these drugs is therefore limited to preanesthetic use.

Eugenol (Clove Oil)

- When administered as a single-bath immersion (dose 350 mg/L) for 15 minutes, eugenol may serve as an effective anesthetic in *X. laevis* frogs for short, minimally invasive surgical procedures or for restraint (Guénette et al. 2007).

Propofol

- When administered as a single-bath immersion for 15 minutes, propofol does *not* appear to be a safe and effective anesthetic for *Xenopus*, because of a narrow dose-effect window, short duration, and shallow level of anesthesia obtained (Guénette et al. 2007).

aseptic surgery

Surgery performed on *Xenopus* should be conducted using aseptic surgical techniques. However, the constant mucous production in the skin of *Xenopus* makes pre-operative prepping and disinfection of the operative area difficult and transient at best, and such prepping will interfere with the mucus's protective mechanisms.

- At a minimum, gross debris should be washed from the incision site using sterile saline. Further disinfection and preparation of the surgical site in *Xenopus* is difficult. Even diluted providone iodine or chlorhexidine solutions will disrupt protective skin flora and the mucous slime coat. Scrubs and detergents should definitely not be used. The surgeon, however, should scrub his or her hands and lower arms with a disinfectant soap, don a mask and sterilized gloves, and use sterile instruments.

- Dry surgical drapes will stick to the slime coat layer on the frog's skin and disrupt its natural protective barrier. Moistened, plastic sterile drapes may be used to keep mucus from getting on instruments and suture material and will keep the skin from drying during surgery. However, wet drapes pose a limited barrier to bacteria.

- Surgical harvest of oocytes is perhaps the most common surgery in laboratory *Xenopus* in which a major body cavity is penetrated. Given the approval of the IACUC, this surgery can be performed multiple times in an anesthetized animal without ill effect, using sterilized instruments and no surgical drapes if the surgeon is competent and quick, and practices all other aspects of aseptic technique. This procedure is described in greater detail in Chapter 5, Experimental Methodology.

- Monitoring records must be kept of all survival surgical procedures in research animals. At the minimum, the loss of righting reflex, the response to toe pinch, and the start and stop time of the surgery should be recorded, as well as the outcome (number of deaths, if any, and number of recoveries without complications). Because gular breathing stops when *Xenopus* are under deep anesthesia and the heart beating beneath the sternum is not always easy to see, further monitoring of vital signs is a challenge without instrumentation.

- Post-surgical closure of skin and muscle layers in *Xenopus* is usually completed with an absorbable monofilament suture material. One group has recently reported that monofilament nylon elicited the least histologic reaction when compared with other materials and therefore may be a good choice for use in amphibian skin (Tuttle et al. 2006).

- Recovery from MS-222 can be relatively quick, within 20 to 30 minutes after removal from the solution. During recovery, *Xenopus* should be partially submerged in fresh tank water, with the head and nares elevated above the water line until they are awake, fully recovered, and able to swim and dive.

analgesics and post-operative care

Heightened awareness for the welfare of earlier-evolved laboratory species has prompted increasing inquiries by institutional animal care committees, investigators, and laboratory animal veterinarians regarding the need for post-surgical analgesics in laboratory *Xenopus*. Because *Xenopus* lack a limbic cortex and their cerebral cortex consists mostly of the olfactory lobes, it has been argued that they do not perceive pain as do mammals. However, *Xenopus* do show a strong avoidance response to noxious nocioceptive stimuli.

Basic research into the mechanisms and regulation of pain in *Rana pipiens* has demonstrated the clinical potential of opioid, alpha2-adrenergic, and non-opioid analgesic agents in amphibians. However, clinical studies using objectively established indices of pain in *Xenopus* and pharmacological studies examining the clinical potential of analgesics in this species have not been conducted. There are currently no pharmacokinetically based recommendations that can be made regarding an efficacious drug or dose of analgesic that can be *safely* administered to *Xenopus*. Limited lethality data suggest that the safety index for analgesics is quite narrow in the semiterrestrial *Rana pipiens*. Analgesic use in the fully aquatic *Xenopus* has the added risk of drowning due to over sedation. Analgesic drugs and drug doses extrapolated from studies conducted in other species should therefore be considered very carefully.

- Some of the analgesics recommended for amphibians include buprenorphine (38 mg/kg SC), butorphanol (0.2–0.4 mg/kg, IM), fentanyl (0.5 mg/kg, SC), meperidine (49 mg/kg, SC), or morphine (38–42 mg/kg, SC) (Gentz 2007). These doses are empirical. Additional analgesic drugs and doses are given in **Table 8**.

- An additional concern for laboratory *Xenopus* is that the effects of repeated treatment with these agents on amphibian oogenesis, oocyte quality, and embryogenesis are unknown.

- Post-operatively, *Xenopus* should not be returned to regular housing after surgery until fully recovered from the general anesthesia and not showing signs of sedation associated with the analgesic (i.e., they can swim and dive). Frogs that are returned to regular housing too soon are at risk of drowning. The animals should be allowed to recover on a slant with the lower body submerged in water and the head and nose elevated. Covering the animal with wet drapes will prevent the skin from drying out during recovery.

- Water in the recovery tanks should be changed regularly, and the animal can be fed a small amount the day after it has fully recovered.

- The post-operative frog should be checked daily to monitor for wound dehiscence and other signs of complications (e.g., lethargy, buoyancy problems, and petechia).

- The practice of adding antibiotics to the frog's water after oocyte harvest surgery is longstanding in the research community. However, there is no documentation that this practice diminishes post-operative infections and the risk of toxicity is very real. Selection of resistant microbial species should also be considered.

euthanasia

- *Euthanasia* means "humane and easy death" and is most easily accomplished in aquatic amphibians via an anesthetic **overdose of MS-222** by prolonged (~30 minutes or more) immersion in a neutrally buffered solution at a concentration of 5–10 g/L water (Gentz 2007).

- Torreilles, McClure, and Green recently reported rapid euthanasia of laboratory *Xenopus* in an immersion of MS-222 as a 5000 mg/L buffered solution, or with an intracoelomic injection with sodium pentobarbital (0.3 ml/frog of Beuthanasia®), or with 2-cm application of 20% Anbesol® to the ventrum (Torreilles et al. 2009).

- Other acceptable physical methods of euthanasia are sedation followed by **decapitation** using guillotine or heavy scissors.

- **Pithing** involves severing the spinal cord at the base of the brain. Pithing is performed by inserting a sharp needle at

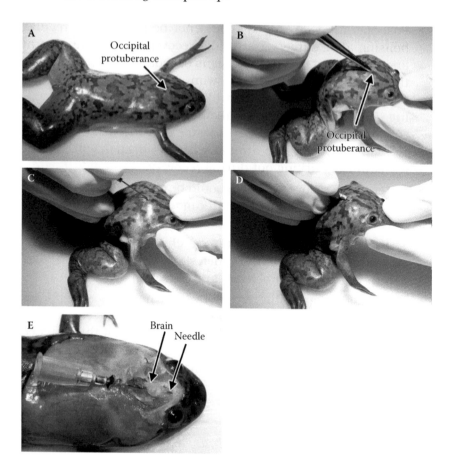

Figure 39 Pithing as a means of euthanasia. This method must be followed by decapitation as a secondary method of euthanasia. This animal was first anesthetized in MS-222. Note the location of the occipital protuberance (A, B). With the animal's head ventroflexed (the position has been exaggerated for the purposes of the photograph), a 22 or 25 g, ½-inch needle can be quickly inserted into the atlanto-occipital joint space (just ventral to the occipital protuberance) and advanced cranially to the hub (C, D). The needle should be moved from side to side to lacerate the brainstem and spinal cord. Note the proximal location of the brain, just behind the eyes (E). (Photo courtesy of Dr. Richard Luong.)

the atlanto-occipital joint and rapidly moving the needle back and forth to transect the cord. The pithing site in frogs is the foramen magnum, and it is identified by a slight midline skin depression caudal to the eyes when the neck is flexed. **(Figure 39)**. Chemical restraint (via MS-222 immersion or administration of the drug IC) *prior* to pithing is the most humane and the least stressful for handler and the animal. However, death may not be immediate unless both the brainstem and spinal cord are completely severed by pithing. Pithing *must* be followed by decapitation as a secondary method of euthanasia.

- Euthanasia by freezing the animal (placing the animal on ice or in a freezer) is painful to the animal and inhumane. It is not an acceptable method. Quick freezing of a *deeply anesthetized* animal is acceptable.

- *Xenopus* can hold their breath and survive prolonged periods of anoxia. Therefore, euthanasia of amphibians using inhalation agents such as CO_2 is not acceptable.

- Other unacceptable methods of euthanizing *Xenopus* include exsanguination, hyperthermia, and electrocution (Wright 2001).

- An overdose of anesthetics such as injectable sodium pentobarbital, 100 mg/kg, IC or SQ into the dorsal lymph sacs (Gentz 2007), immersion in buffered benzocaine hydrochloride (at concentrations >250–300 g/L) bath, or administration of the buffered solution by IC or SQ injection into the dorsal lymph sac are alternative euthanasia methods.

references

Bernstein, P. S., K. B. Digre, and D. J. Creel. 1997. Retinal toxicity associated with occupational exposure to the fish anesthetic MS-222. *Am. J. Opthalmol.* 124(6):843–44.

Blumer, C., D. R. Zimmermann, R. Weilenmann, L. Vaughan, and A. Pospischil. 2007. Chlamydiae in free-ranging and captive frogs in Switzerland. *Vet. Pathol.* 44:144–50.

Bodetti, T. J., E. Jacobson, C. Wan et al. 2002. Molecular evidence to support the expansion of the host range of *Chlamydophila pneumoniae* to include reptiles as well as humans, horses, koalas and amphibians. *Syst. Appl. Microbiol.* 25:146–52.

Bollinger, T. K., J. Mao, D. Schock, R. M. Brigham, and V. G. Chinchar. 1999. Pathology, isolation, and preliminary molecular characterization of a novel iridovirus from tiger salamanders in Saskatchewan. *J. Wildl. Dis.* 35:413–29.

Brayton, C. 1992. Wasting disease associated with cutaneous and renal nematodes, in commercially obtained *Xenopus laevis*. *Ann. N. Y. Acad. Sci.* 653:197–201.

Brown, H. H. K., H. K. Tyler, and T. A. Mousseau. 2004. Orajel as an amphibian anesthetic: refining the technique. *Herpetol. Rev.* 35(3):252.

Bruce, H. M. and A. S. Parkes. 1950. Rickets and osteoporosis in *Xenopus laevis*. *J. Endocrinol.* 7:64–81.

Chai, N., L. Deforges, W. Sougakoff et al. 2006. *Mycobacterium szulgai* infection in a captive population of African clawed frogs (*Xenopus tropicalis*). *J. Zoo Wildlife Med.* 37:55–58.

Chen, M. H. and C. A. Combs. 1999. An alternative anesthesia for amphibians: ventral application of benzocaine. *Herpetol. Rev.* 30(1):34.

Clothier, R. H. and M. Balls. 1973. Mycobacteria and lymphoreticular tumours in *Xenopus laevis*, the South African clawed toad. I. Isolation, characterization and pathogenicity for *Xenopus* of *M. marinum* isolated from lymphoreticular tumour cells. *Oncology* 28:445–57.

Colt, J. 1986. Gas supersaturation—impact on the design and operation of aquatic systems. *Aquac. Eng.* 5:49–85.

Colt, J., K. Orwicz, and D. Brooks. 1984. Gas bubble disease in the African clawed frog, *Xenopus laevis*. *J. Herpetol.* 18(2):131–37.

Colt, J. E. 1983. The computation and reporting of dissolved gas levels. *Water Res.* 17(8):841–49.

Colt, J. E., G. Bouck, and L. E. Fidler. 1986. Review of current literature and research on gas supersaturation and gas bubble trauma. U.S. Dept. of Energy, Bonneville Power Admin., Div. of Fish and Wildlife. Special Publication No.1.

Crawshaw, G. J. 1992. The role of disease in amphibian decline. In *Declines in Canadian amphibian populations: designing a national monitoring strategy*, Occasional Paper no. 76 edition, ed. C. A. Bishop, and K. E. Pettit, 60–62. Ottawa: Canadian Wildlife Service.

Cunningham, A. A., A. W. Sainsbury, and J. E. Cooper. 1996. Diagnosis and treatment of a parasitic dermatitis in a laboratory colony of African clawed frogs (*Xenopus laevis*). *Vet. Rec.* 138:640–42.

Densmore, C. L. and D. E. Green. 2007. Diseases of amphibians. *ILAR J.* 48:235–54.

Downes, H. 1995. Tricaine anesthesia in amphibia: A review. *Bull. Assoc. Rept. Amphib. Vet.* 5(2):11–16.

Duellman, W. E. and L. Trueb. 1994. Reproductive strategies. In *Biology of Amphibians*, ed. W. E. Duellman and L. Trueb, 13–50. Baltimore: Johns Hopkins University Press.

Elkan, E. 1960. Some interesting pathological cases in amphibians. *Proc. Zool. Soc. Lond.* 34:375–96.

Elsner, H. A., H. H. Honck, F. Willmann, H. J. Kreienkamp, and F. Iglauer. 2000. Poor quality of oocytes from *Xenopus laevis* used in laboratory experiments: prevention by use of antiseptic surgical technique and antibiotic supplementation. *Comp. Med.* 50(2):206–11.

Emerson, H. and C. Norris. 1905. "Red-leg," an infectious disease of frogs. *J. Exp. Med.* 7:32–58.

Ford, T. R., D. L. Dillehay, and D. M. Mook. 2004. Cutaneous acariasis in the African clawed frog (*Xenopus laevis*). *Comp. Med.* 54(6):713–17.

Gentz, E. J. 2007. Medicine and surgery of amphibians. *ILAR J.* 48:255–59.

Godfrey, D., H. Williamson, J. Silverman, and P. L. Small. 2007. Newly identified *Mycobacterium* species in a *Xenopus laevis* colony. *Comp. Med.* 57:97–104.

Green, D. E. 2001. Pathology of amphibia. In *Amphibian medicine and captive husbandry*, ed. K. M. Wright and B. R. Whitaker, 401–85. Malabar, FL: Krieger Publishing Company.

Green, D. E. and J. C. Harshbarger. 2001. Spontaneous neoplasia in amphibia. In *Amphibian medicine and captive husbandry*, ed. K. M. Wright and B. R. Whitaker, 335–400. Malabar, FL: Krieger Publishing Company.

Green, S. L. 2002. Factors affecting oogenesis in the South African clawed frog (*Xenopus laevis*). *Comp. Med.* 52:307–12.

Green, S. L., D. M. Bouley, C. A. Josling, and R. Fayer. 2003. Cryptosporidiosis associated with emaciation and proliferative gastritis in a laboratory South African clawed frog (*Xenopus laevis*). *Comp. Med.* 53(1):81–84.

Green, S. L., D. M. Bouley, R. J. Tolwani et al. 1999. Identification and management of an outbreak of *Flavobacterium meningosepticum* infection in a colony of South African clawed frogs (*Xenopus laevis*). *J. Am. Vet. Med. Assoc.* 214:1833.

Green, S. L., B. D. Lifland, D. M. Bouley et al. 2000. Disease attributed to *Mycobacterium chelonae* in South African clawed frogs (*Xenopus laevis*). *Comp. Med.* 50:675–79.

Green, S. L., R. C. Moorhead, and D. M. Bouley. 2003. Thermal shock in a colony of South African clawed frogs (*Xenopus laevis*). *Vet. Rec.* 152:336–37.

Green, S. L., J. Parker, C. Davis, and D. M. Bouley. 2007. Ovarian hyperstimulation syndrome in gonadotropin-treated laboratory South African clawed frogs (*Xenopus laevis*). *J. Am. Assoc. Lab. Anim. Sci.* 46(3):64–67.

Guénette, S. A., P. Hélie, F. Beaudry, and P. Vachon. 2007. Eugenol for anesthesia of African clawed frogs (*Xenopus laevis*). *Vet. Anaesth. Analg.* 34(3):164–70.

Hadfield, CA. and B. R. Whitaker. 2005. Amphibian emergency medicine and care. *Semin. Avian Exot. Pet Med.* 1 4:79–89.

Hall, A. H., S. J. Newman, L. Craig, C. Carter, J. Czarra, and J. Paige Brown, 2009. Diagnosis of *Aeromonas hydrophila* and *Batrachochrtrium dendrobatidis* in an African clawed frog (*Xenopus laevis*). *J Comp Med*, in press.

Howerth, E. W. 1984. Pathology of naturally occurring chlamydiosis in African clawed frogs (*Xenopus laevis*). *Vet. Pathol.* 21:28–32.

Howerth, E. W. and J. M. Pletcher. 1986. Diagnostic exercise: death of African clawed frogs. *Lab. Anim. Sci.* 36(3):286–87.

Iglauer, F., F. Willmann, G. Hilken, E. Huisinga, and J. Dimigen. 1997. Anthelmintic treatment to eradicate cutaneous capillariasis in a colony of South African clawed frogs (*Xenopus laevis*). *Lab. Anim. Sci.* 47:477–82.

Jacobson, E., F. Origgi, D. Heard, and C. Detrisac. 2002. Immunohistochemical staining of chlamydial antigen in emerald tree boas (*Corallus caninus*). *Lab. Anim. Sci.* 14:487–94.

Kaiser, H. and D. M. Green. 2001. Keeping the frogs still: Orajel® is a safe anesthetic in amphibian photography. *Herpetol. Rev.* 32:93–94.

Kriger, K. M., H. B. Hines, A. D. Hyatt, D. G. Boyle, and J. M. Hero. 2006. Techniques for detecting chytridiomycosis in wild frogs: Comparing histology with real-time Taqman PCR. *Dis. Aquat. Organ* 71:141–48.

Lucke, B. 1934. A neoplastic disease of the kidney of the frog, *Rana pipiens. Am. J. Cancer* 20:352–79.

Machin, K. L. 1999. Amphibian pain and analgesia. *J. Zoo Wildlife Med* 30:2–10.

Machin, K. L. 2001. Fish, amphibian, and reptile analgesia. *Vet. Clin. North Am. Exot. Anim. Pract.* 4:19–33.

Manuel, M. J., D. L. Miller, K. S. Frazier, and M. E. Hines, II. 2002. Bacterial pathogens isolated from cultured bullfrogs (*Rana catesbeiana*). *J. Vet. Diagn. Invest.* 14:431–33.

Mazzoni, R., A. A. Cunningham, P. Daszak et al. 2003. Emerging pathogen of wild amphibians in farms (*Rana catesbeiana*) farmed for international trade. *Emerg. Infect. Dis.* 9:995–98.

McLaren, I. A. 1965. Temperature and frog eggs. A reconsideration of metabolic control. *J. Gen. Physiol.* 48(6):1071–79.

Newcomer, C. E., M. R. Anver, J. L. Simmons, B. W. Wilcke, Jr., and G. W. Nace. 1982. Spontaneous and experimental infections of *Xenopus laevis* with *Chlamydia psittaci*. *Lab. Anim. Sci.* 32(6):680–86.

Olson, M. E., S. Gard, M. Brown, R. Hampton, and D. W. Morck. 1992. *Flavobacterium indologenes* infection in leopard frogs. *J. Am. Vet. Med. Assoc.* 201:1766–70.

Parker, J. M., I. Mikaelian, N. Hahn, and H. E. Diggs. 2002. Clinical diagnosis and treatment of epidermal chytridiomycosis in African clawed frogs (*Xenopus tropicalis*). *Comp. Med.* 52:265–68.

Pessier, A. P. 2002. An overview of amphibian skin disease. *Semin. Avian Exot. Pet Med.* 11:162–74.

Pessier, A. P., D. K. Nichols, J. E. Longcore, and M. S. Fuller. 1999. Cutaneous chytridiomycosis in poison dart frogs (*Dendrobates* spp.) and White's tree frogs (*Litoria caerulea*). *J. Vet. Diagn. Invest.* 11:194–99.

Poynton, S. L. and B. R. Whitaker. 2001. Protozoa and metazoa infecting amphibians. In *Amphibian medicine and captive husbandry*, ed. K. M. Wright and B. R. Whitaker, 193–222. Malabar, FL: Krieger Publishing Company.

Pritchett, K. R. and G. E. Sanders. 2007. Epistylididae ectoparasites in a colony of African clawed frogs (*Xenopus laevis*). *J. Am. Assoc. Lab. Anim. Sci.* 46:86–91.

Reed, K. D., G. R. Ruth, J. A. Meyer, and S. K. Shukla. 2000. *Chlamydia pneumoniae* infection in a breeding colony of African clawed frogs (*Xenopus tropicalis*). *Emerg. Infect. Dis.* 6:196–99.

Robert, J., L. Abramowitz, and H. D. Morales. 2007. *Xenopus laevis*: a possible vector of *Ranavirus* infection? *J. Wildlife Dis.* 43(4):645–52.

Rollins-Smith, L. A., J. P. Ramsey, L. K. Reinert, D. C. Woodhams, L. J. Livo, and C. Carey. 2009. Immune defenses of *Xenopus laevis* against *Batrachochytrium dendrobatidis*. *Front Biosci (Schol Ed)* S1: 68–91.

Ruble, G., I. K. Berzins, and D. L. Huso. 1995. Diagnostic exercise: anorexia, wasting, and death in South African clawed frogs. *Lab. Anim. Sci.* 45:592–94.

Sánchez-Morgado, J., A. Gallagher, and L. K. Johnson. 2009. *Mycobacterium gordonae* infection in a colony of African clawed frogs (*Xenopus tropicalis*). *Lab. Anim.* Feb 23.

Schuytema, G. S. and A. V. Nebeker. 1999. Effects of ammonium nitrate, sodium nitrate, and urea on red-legged frogs, Pacific treefrogs, and African clawed frogs. *Bull. Environ. Contam. Toxicol.* 63(3):357–64.

Schwabacher, H. 1959. A strain of *Mycobacterium* isolated from skin lesions of a cold-blooded animal, *Xenopus laevis*, and its relation to atypical acid-fast bacilli occurring in man. *J. Hyg. (Lond).* 57:57–67.

Stephens, L. C., D. M. Cromeens, V. W. Robbins, P. C. Stromberg, and J. H. Jardine. 1987. Epidermal capillariasis in South African clawed frogs (*Xenopus laevis*). *Lab. Anim. Sci.* 37:341–44.

Stevens, C. W., D. N. MacIver, and L. C. Newman. 2001. Testing and comparison of non-opioid analgesics in amphibians. *Contemp. Top. Lab. Anim. Sci.* 40(4):23–27.

Suykerbuyk, P., K. Vleminckx, F. Pasmans et al. 2007. *Mycobacterium liflandii* infection in European colony of *Silurana tropicalis*. *Emerg. Infect. Dis.* 13:743–46.

Taylor, S. K. 2001a. Bacterial diseases. In *Amphibian medicine and captive husbandry*, ed. K. M. Wright and B. R. Whitaker, 159–180. Malabar, FL: Krieger Publishing Company.

Taylor, S. K. 2001b. Mycoses. In *Amphibian medicine and captive husbandry*, ed. K. M. Wright and B. R. Whitaker, 181–192. Malabar, FL: Krieger Publishing Company.

Tinsley, R. C. 1996. Parasites of *Xenopus*. In *The biology of Xenopus*, ed. R. C. Tinsley and H. R. Kobel, 233–61. Oxford: Oxford University Press.

Torreilles, S. L. and S. L. Green. 2007. Refuge cover decreases the incidence of bite wounds in laboratory South African clawed frogs (*Xenopus laevis*). *J. Am. Assoc. Lab. Anim. Sci.* 46:33–36.

Torreilles, S. L., McClure, D. E., and Green, S. L. 2009. Euthanasia methods for laboratory South African clawed frogs (*Xenopus laevis*): a comparative study. *J. Am. Assoc. Lab. Anim. Sci.* In press.

Trott, K. A., B. A. Stacy, B. D. Lifland et al. 2004. Characterization of a *Mycobacterium ulcerans*-like infection in a colony of African tropical clawed frogs (*Xenopus tropicalis*). *Comp. Med.* 54:309–17.

Tuttle, A. D., J. M. Law, C. A. Harris, G. A. Lewbart, and S. B. Harvey. 2006. Evaluation of the gross and histologic reactions to five commonly used suture materials in the skin of the African clawed frog (*Xenopus laevis*). *J. Am. Assoc. Lab. Anim. Sci.* 45(6):22–26.

Weldon, C., L. H. du Preez, D. D. Hyatt, R. Muller, and R. Speare. 2004. Origin of the amphibian cytrid fungus. *Emerg Infect Dis* 10:2100–2105.

Whitaker, B. R. 2001. Water quality. In *Amphibian medicine and captive husbandry*, ed. K. M. Wright and B. R. Whitaker, 147–57. Malabar, FL: Krieger Publishing Company.

Wilcke, B. W., Jr., C. E. Newcomer, M. R. Anver, J. L. Simmons, and G. W. Nace. 1983. Isolation of *Chlamydia psittaci* from naturally infected African clawed frogs (*Xenopus laevis*). *Infect. Immun.* 41(2):789–94.

Wright, K. M. 2001. Restraint techniques and euthanasia. In *Amphibian medicine and captive husbandry*, ed. K. M. Wright, and B. R. Whitaker, 111–22. Malabar, FL: Krieger Publishing Company.

Wright, K. M. 2006. Overview of amphibian medicine. In *Reptile medicine and surgery*, 2nd edition, ed. D. R. Mader, 941–71. St. Louis: Saunders, Elsevier.

Wright, K. M. 2009. Amphibians. In *Exotic animal formulary*, 2nd edition, ed. J. W. Carpenter, T. Y. Mashima, and D. J. Rupiper, 23–38. Philadelphia: W.B. Saunders.

experimental methodology

catching and handling *Xenopus*

- A net should be used to scoop *Xenopus* out of the water. Cover the net opening with your hand to prevent the frog from jumping out during transport. The net avoids damage to the protective mucous layer on the frog's skin and is less traumatic than hand contact with the frog.
- Wear moistened, powder-free nonlatex gloves when handling *Xenopus*. Wetting the gloves with tank water protects the slime coat layer on the frog and will not harm the animal. Gloves protect the handler from potential zoonotic infections.
- The slime coat on *Xenopus* skin is very slippery. They can be difficult to restrain; hold the frog as shown in **Figure 40**. The frog's hind legs are between the handler's index finger and thumb and the frog's dorsum is against the handler's palm. Covering the frog's eyes during restraint tends to calm the animal. It may be necessary to use two hands to form a cage around the animal. Injections are best administered by placing the frog on a table surface and restraining the frog with one hand while administering the injection with the other (see below).

Figure 40 Restraining a laboratory *Xenopus*. Note the frog's hind legs are between the handler's thumb and index finger and the animal's eyes are covered by the handler's palm.

compound administration techniques

- ***Subcutaneous injections*** are given to frogs via the dorsal lymph sac, into the subcutaneous space close to the location where the posterior lymph hearts are located.
- The needle gauge should be the smallest possible, 22 to 28 with a ½- to ¾-inch length. The volume should be less than 0.5 ml. With the needle placed almost parallel to the frog's body (the subcutaneous space is shallow, just beneath the skin) and just medial to the lateral line, penetrate the skin with the bevel of the needle facing upward (**Figure 41**). A subcutaneous injection into the dorsal lymph sac can be viewed at http://www.jove.com/index/Details.stp?ID=890 (Cross and Powers 2008).
- The quadriceps muscles can be used for ***intramuscular injections***. Using a 22-g needle (¾ inch), the volume that can easily be administered intramuscularly is usually 0.5 ml or less at any one site, depending on the size of the frog (**Figure 42**).
- ***Intracoelomic injections*** can be given and a slightly larger volume administered (1–2 ml, depending on the size of the frog). For intracoelomic injections, the frog can be held in dorsal recumbency with its head directed downward so that

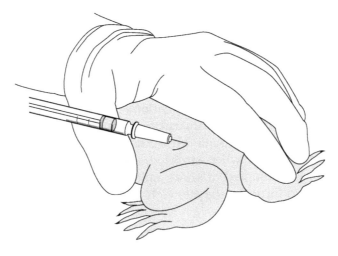

Figure 41 Subcutaneous injection into the dorsal lymph sac of a laboratory *Xenopus*.

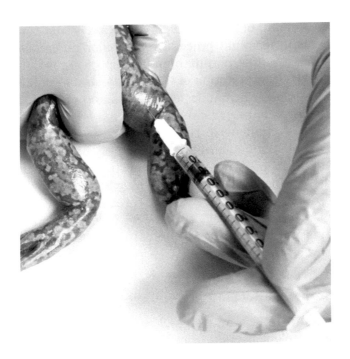

Figure 42 Intramuscular injection into the quadriceps muscle of a laboratory *Xenopus*.

viscera fall away from the injection entry site and the needle inserted in the lower quadrant of the coelomic cavity. Gently aspirate the syringe and check for the presence of blood in the syringe. If blood is present, the needle may be placed in the spleen. Withdraw and redirect the needle.

- In *Xenopus*, injection of medications or hormones is preferable to **per os** or **oral administration**. Amphibians absorb drugs poorly from the stomach, and *Xenopus* will regurgitate noxious materials or when stressed. If the oral route of administration is attempted, a short, soft, small-gauge catheter can be used to intubate the stomach.

blood sample collection and interpretation

Despite the long history of laboratory *Xenopus* in research, hematological and clinical chemistry evaluations have not been used extensively as a part of their clinical care. There are a number of reasons for this: 1) Many extrinsic and intrinsic factors (age, gender, water temperature, time of the year, etc.) influence the results, 2) to obtain an adequate amount of blood, sample collection is best performed via cardiocentesis, a terminal procedure, 3) major differences between mammals and aquatic, non-mammalian species pose a significant challenge in the interpretation of the results, and 4) published reference range values for hematological indices and clinical chemistries for *Xenopus* have not been established.

Despite these challenges, the clinical assessment of laboratory *Xenopus* can be augmented by evaluation of the clinical chemistry and hematological profiles. As the use of this species continues and the model is refined, the utility of these diagnostics will improve. Blood collection techniques, sample processing, a summary of the hematological characteristics unique to *Xenopus*, and a commentary on interpretation of blood clinical chemistry results are presented below.

Blood Sample Collection

- *Xenopus* are tailless and tongueless aquatic amphibians. Thus, blood sample collection from the tail vein or lingual vein, as performed in other frog species, is not applicable. The femoral vein is small and difficult to isolate percutaneously. Neither do *Xenopus* have a prominent midline abdominal vein, as do bullfrogs and salamanders. Amputation of a digit will yield a few drops of blood, but the sample will be contaminated with lymph and tissue debris. Thus blood sample collection from *Xenopus* is most easily accomplished via cardiocentesis.

- Cardiocentesis should be performed in a *fully* anesthetized animal and only in animals that are not going to be recovered. A secondary method of euthanasia is recommended to ensure death after the procedure.

- The anticoagulant of choice for collection instruments and sample storage is lithium heparin. Ethylenediamine-tetra-acetic acid (EDTA) is not recommended because it tends to lyse *Xenopus* red blood cells (RBCs) (Allender and Fry 2008).

- The animal should be sedated via tricaine methansulfonate (MS-222) immersion or intracoelomic injection, placed in dorsal recumbency with cardiac impulse visualized just beneath the sternum.

- Percutaneous cardiocentesis can be performed with practice, but for the best access and visibility, the coelomic and thoracic cavity should be opened and the heart viewed directly, as shown in **Figure 43**.

- The hub of a lithium-heparin pre-treated microhematocrit tube can be inserted directly into the ventricle and the

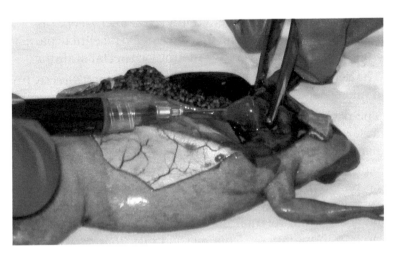

Figure 43 Blood sample collection via terminal pericardiocentesis of a laboratory *Xenopus*.

lood allowed to fill the tube as the heart continues to beat. Alternatively, a 22-, 25-, or 26-gauge needle attached to a 1- or 3-ml lithium-heparin pretreated syringe can be inserted into the ventricle and blood collected via gentle aspiration on the attached lithium-heparin pretreated syringe. The *Xenopus* pericardium is resistant to puncturing, and the heart should be stabilized with forceps when the needle or hematocrit tube is inserted.

- The blood volume of *Xenopus* is ~13.4% (Wright 2001). Using the methods described here, 2 to 7 ml of blood from a ~110-gram adult female frog can be obtained.

- Digital amputation (of a hind toe) has been used as a means of collecting blood samples from *Xenopus* in the field and in some laboratory environments; however, this method is painful and puts the frog at risk for infection. In addition, as stated above, such samples are often contaminated with tissue debris and extracellular fluids such that the results are uninterpretable.

Processing the Samples

- *Xenopus* RBCs are quite fragile, so samples should be handled gently to yield optimal results.

- Small volumes of whole blood should be immediately placed on a glass slide and viewed under a microscope as a wet mount to evaluate for the presence of extracellular parasites, and then prepared as a dry mount for special staining.

- The hematocrit tube itself can also be examined directly under a microscope for the presence of flagellated or other parasites.

- The packed cell volume (PCV), percent buffy coat, and icteric index can be obtained from the remaining sample in the hematocrit tube. *Xenopus* plasma color ranges from a clear pale yellow to gold. After centrifugation, plasma can then be removed to determine total solids or total protein using a refractometer. However, refractometers are calibrated based on the normal relation between refractive index and plasma protein in mammals. This method may therefore not be reliable in *Xenopus* (Allender and Fry 2008). To obtain an accurate plasma protein measurement, the spectrophotometric (biruret) method is required.

- The air-dried whole blood mount glass slides should be stained with Wright's solution (Camco Quik Stain, American Scientific Products, McGaw Park, Illinois) and evaluated for parasites, erythrocytes, and platelets, the leukocyte differential determined, and the number of leukocytes counted per high-power field.

- A Gram stain of the blood film may be useful for identifying bacteria if bacteremia is suspected. If toxic or phagocytic monocytes are noted and there is sufficient sample available, the total white blood cell count should then be determined.

- As in birds, reptiles, and other amphibians, *Xenopus* have nucleated RBCs and thrombocytes (the amphibian equivalent of the mammalian platelet). Thus, the automated cell counting systems typically used to determine total cell counts for mammalian species cannot be used. Examples of tests that are not valid in amphibians using conventional automated methods include cell counts, hemoglobin concentration, mean cell volume (MCV), mean cell hemoglobin (MCH), and mean cell hemoglobin concentration (MCHC). Instead, complete blood counts, or total white blood cell counts per mm^3 using a hemocytometer, an erythrocyte diluting pipette, and Natt-Herrick's solution (which stains all cells) must be performed (Wright 2001). This technique allows the diagnostician to hand-count leukocytes, erythrocytes, and thrombocytes.

- Serum chemistry parameters can be evaluated on dry film analyzers (such as the Kodak Katachem DT60 Analyzer, Eastman Kodak Co., Rochester, NY). The small volume (10 µl) of sample needed for analysis by these machines allows processing of a 0.5 ml plasma sample to analyze up to five different chemistries.

Xenopus hemocytology: characteristics

Xenopus blood cells tend to be pleomorphic, with great variation in their staining characteristics compared to mammalian and avian blood cells. The recognition of the various blood cell types circulating in *Xenopus* requires experience and special stains for distinguishing cell types. General amphibian hematology is reviewed in great detail by Allender and Fry (2008) and by Wright (2001). Hadji-Azimi has also published a review of *Xenopus* hemocytology, and although

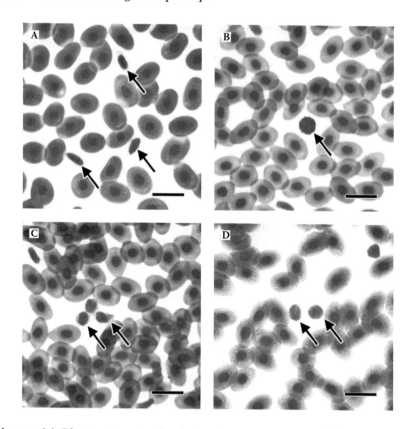

Figure 44 Photomicrograph of blood smears prepared from a healthy female laboratory *Xenopus* showing nucleated erythrocytes and (A) three thrombocytes, (B) a heterophil, (C) three azurophils (the amphibian equivalent of mammalian monocytes), and (D) two lymphocytes. Bar = 20 microns. Modified Wright-Giemsa stain. (Photo courtesy of Dr. Richard Luong).

the number of frogs sampled for the study is low (seven) and the age, source, and gender not specified, the report features useful microphotographs of the different cell types of this species (Hadji-Azimi et al. 1987). The photomicrographs in **Figure 44** are blood smears prepared from a healthy female laboratory *Xenopus*. Some of the most unique features of *Xenopus* hematology are summarized below.

- *Xenopus* **erythrocytes** are large, varying from 10 to 70 microns in diameter, elliptical, eosinophilic, and nucleated. Small, round, basophilic immature erythroblasts are sometimes observed in the circulation of healthy *Xenopus* and not necessarily indicative of disease.

- **Thrombocytes (Figure 44 A)** are smaller than the RBCs, also nucleated, but are elliptical-shaped cells. They are prone to fragmentation during processing. Thrombocytes in *Xenopus* can be difficult to distinguish from lymphocytes. *Xenopus* lymphocytes rarely aggregate as do thrombocytes.

- The normal ratio of leukocytes to erythrocytes in amphibian blood is generally between 1:20 and 1:70 (Duellman and Trueb 1986). Normal total and differential leukocyte counts for *Xenopus* are not known, and few reports describe leukogram changes in disease.

- Because different types of amphibian white blood cells can be difficult to identify definitively without special stains, the term **heterophil (Figure 44 B)** is frequently used when describing amphibian phagocytic white blood cells.

- Amphibian **granulocytes** are named according to their morphological description when prepared with Wright-Giemsa stain and are named accordingly: large eosinophilic granulocytes, heterophilic granulocytes (a small eosinophilic granule leukocyte), basophilic granulocyte, neutrophilic granulocyte, and azurophilic granulocyte.

- The presence of **eosinophils** does not necessarily indicate parasitism or allergic response in *Xenopus* (these cells are often observed in healthy animals). While eosinophils may have some effect against parasites in other amphibians, their function in *Xenopus* is not clear.

- **Basophils** are typically less than 1% of the circulating granulocytes in most amphibian species (Cannon and Cannon 1979).

- **Mast cells** can be difficult to distinguish from basophils, but the mast cell is generally smaller, contains finer granules, and has an oval, more rounded nucleus and more cytoplasm than the basophil.

- **Azurophils (Figure 44 C)** have a light gray-blue cytoplasm and are thought to be of the same cell type as the mammalian monocyte or neutrophil in reptiles.

- **Lymphocytes (Figure 44 D)** are round with a small amount of cytoplasm and a round, bilobed nucleus. It can be difficult to distinguish lymphocytes from the thrombocytes, monocytes, immature erythroid cells, and plasma cells. Lymphocytes are the smallest leukocyte, about half the size of a mature erythrocyte. Monocytes are almost always peroxidase-positive, while lymphocytes are peroxidase-negative, and thrombocytes tend to aggregate (lymphocytes rarely aggregate).

- Plasma cells are extremely uncommon.

- The immune system of *Xenopus* has been studied in great detail (for review, see Du Pasuier et al. 1996; Horton et al. 1996). Lymphocytes have been analyzed cytochemically, are known to produce immunoglobulin isotypes (IgM, and IgX, an IgG-like isotype) and are involved in antibody-dependent, immune-mediated cytotoxicity.

- The spleen is the primary site of erythopoiesis in adult *Xenopus*, though the kidney, bone marrow, and liver also have a role.

interpretation of the hemogram and serum clinical chemistries

Hematological values expressed as a percentage of the total numbers of cells for a small number (fewer than 20 animals in each) of *Xenopus* have been described (Hadji-Azimi et al. 1987; Krishnamoorthy and Shakunthala 1974; Sonnen 1970), and values from more recent publications have been reported (Chu et al. 2006; Green et al. 2003; Wright 2001); however, the information is of limited value given the low number of animals, the variation in the animals' age, gender, source, and housing conditions, and the single sampling. Reports that compare values between a normal population of healthy *Xenopus* and those with disease are also rare (Green et al. 2003). However, as a diagnostic aid the hemogram is still quite useful when the morphology of the erythrocytes and leukocytes reveals the presence or absence of viral inclusions, parasites, bacteria, or toxic neutrophils. To the author's knowledge there are no published reports describing *Xenopus* plasma clinical chemistry values, either as reference ranges or as samples collected from diseased animals. Comparison of the hemograms and serum clinical chemistries from healthy frogs with those collected from sick frogs in the same population may be helpful.

egg/oocyte collection

Egg and oocyte collection from adult female laboratory *Xenopus* is one of the most common experimental methodologies. Veterinarians,

veterinary nurses, and animal caretakers may not routinely per-
form egg collection for a research lab; they should, however, become
familiar with the procedure and have a basic understanding of
what is involved. There are numerous sources describing methods for egg and oocyte collection (Browne and Zippel 2007; Gentz
2007; Sive et al. 2000) and for the generation of transgenic frogs
(Loeber et al., 2009). In a laboratory environment, *Xenopus* are
hormonally "primed" (given an injection of pregnant mares' serum
gonadotropin—PMSG), which triggers the frogs' oocytes to mature
to stage 6, and induced to ovulate for the purpose of providing a
steady supply of material for research. After priming and ovulation, eggs and oocytes can be collected in several ways:

1. *Natural mating* (placing the primed male and female together).
   The female will spawn. Some eggs obtained this way will
   be fertilized.

2. *Stripping, or "milking," the eggs from the female* by gentle palpation or stroking downward on the frog's ventrum. This
   method will yield eggs and oocytes at various stages. Mature
   eggs can then be fertilized *in vitro*.

3. *Collecting the eggs from the female via surgical laparotomy.*
   This procedure is done under general anesthesia and causes
   little harm to the animal. Surgical harvesting is a means of
   collecting the stage I–VI oocytes required for specific biological experiments.

Simple protocols for each of the above are:

1. **Hormonal Priming, Induction of Ovulation and Egg
   Laying**
   - There are many protocols for priming and induction of
     ovulation. An example is given below. More protocols can
     be found in Chapter 6.
   - Female *X. laevis* should receive a typical priming dose
     of 10 IU HCG, given via IC or lymph sac injection 12 to
     72 hours before an ovulatory dose of 100 to 300 IU. For a
     female kept at 20° to 23°C (room temperature), egg laying
     should begin about 10 to 12 hours later. Ovulated females
     should be rested a minimum of 1 to 3 months between
     cycles of priming and egg laying. Six months is ideal. Nine
     months is too long (egg production diminishes).

2. **Egg Laying and Collection after Natural Mating**

- For spawning, males and females are paired and housed together in about 6 inches of deep solution of 20 NM NaCl. Ovulation, mating, and subsequent egg laying can be induced using a dose of 200 IU HCG for females and 100 IU HCG for males, given in the evening, resulting in mating and egg laying by the morning.

- The mating pair should be placed in a bucket and on an elevated mesh grid so that eggs can fall through and not be damaged or eaten by the mating pair. Eggs should be promptly collected and can be stored in the short term in an egg-laying buffer (for recipes, see Chapter 6).

3. **Milking, or Stripping, the Eggs from *Xenopus***

- Eggs may be collected from a primed female by stripping, or "**milking**," the eggs from her by applying pressure to the lower coelomic cavity, stroking downward and laterally on the ventrum to express the eggs through the cloaca. This mimics the physical stimulation of amplexus. Eggs can be milked into a sterile Petri dish containing 1X MBS high-salt solution.

- After 1 or 2 minutes of stroking or milking, the frog will begin to lay eggs. Eggs should not be squeezed out of the frog, and extensive stroking pressure is not required. Trauma (bruising and hemorrhage) to the body wall musculature and rupture of internal organs (liver and spleen) have been observed in *Xenopus* when eggs were expressed with too much pressure.

- Eggs can be milked from the primed frog every hour or so for the first few hours. A maximum of 4 to 6 collections can be expected from one frog over a 24-hour collection period. Eggs can be stripped from the same female after a recovery period of 3 to 6 months.

4. **Surgical Harvest of Oocytes**

Surgical harvesting of stages I–IV oocytes requires general anesthesia, a laparotomy, and exteriorization of the ovarian mass, followed by removal of a small portion of the ovary. The operation, when performed correctly, causes little harm to the frog and can be performed several times in the course of the frog's life without ill effect, but the operation must be performed on an anesthetized frog using aseptic technique.

The Guide for the Care and Use of Laboratory Animals discourages multiple survival surgeries and requires adequate justification and IACUC approval before the procedures are performed. Many institutions limit the number of oocyte harvesting surgeries that can be performed on any one frog to six surgeries (with the sixth being terminal). Subsequent surgeries should be performed on the contra lateral side. A 3- to 6-month rest period between surgeries is recommended. Females that are not used for egg or oocyte harvest (via surgical harvest, natural mating, or milking) for more than 3 to 6 months tend to produce an increase in necrotic eggs and poorer quality oocytes.

An example of the surgical approach to egg harvest from *Xenopus* is given below (adapted from guidelines of the NIH Office of Animal Care and Use (Animal Research Advisory Committee 2005).

Surgical Laparotomy for Egg/Oocyte Harvest from *Xenopus*

- Place the MS-222 anesthetized frog in dorsal recumbency. Artificial slime, such as Shield-X® (Aquatronic, Oxnard, CA) water may be applied to the frog's skin to keep it moist. Gross debris is washed away from the ventrum using a squirt bottle filled with sterile saline. The frog can be covered with sterile saline-moistened plastic surgical drapes, leaving the small area where the skin incision is made uncovered.

- The skin and abdominal muscle incisions should be made in two stages. A diagonal 1 cm or less incision is made through the skin in the lower quadrant of the abdomen with a pointed #11 scalpel blade.

- After incising the skin, lift the muscle layer with surgical forceps and make the incision in the tented muscle. Avoid wounding internal organs or transecting lymph hearts or blood vessels.

- Grasp the ovary with forceps and exteriorize a portion of the oocyte mass. The exteriorized portion of the ovarian mass should not come into contact with the frog's skin (the skin is not sterile). The ovarian mass can be laid gently on the waterproof sterile drape used to cover the animal's ventrum and the biopsy sample collected or snipped from the forceps-suspended ovarian mass.

- The desired numbers of oocytes are excised, and the remainder of the ovarian mass should be carefully replaced in the coelomic cavity. There should be no bleeding associated with the excision.

- The incision should be closed in two layers with absorbable gut (5-0 chromic gut) or PDS for the muscle layer and nylon (non-wicking) suture for the skin closure (5-0 or 6-0 monofilament nylon or PDS). A simple interrupted or horizontal mattress pattern of suturing, rather than continuous, is recommended for skin closure.

- The removal of absorbable sutures is not needed as they gradually dissolve over time. The animals should be monitored for the first week after surgery, and non-absorbable sutures should be removed in 10 to 14 days.

- Some recommend the post-operative use of analgesics. However, pain assessment in *Xenopus* post-oocyte laparotomy is difficult. Also, since their efficacy is difficult to determine, the use of analgesics is not without risk of oversedation and drowning. If the oocyte-harvesting surgeon is competent and quick, *Xenopus* recovered from the general anesthesia can be returned to normal housing in 24 to 48 hours.

raising *Xenopus* tadpoles

Research laboratories requiring a steady supply of eggs and oocytes collected from female *Xenopus laevis* generally find it most cost effective to purchase commercially bred and reared adult frogs from reputable suppliers, rather than breeding and raising their own frogs. However, researchers who use the species to study vertebrate embryology and development do raise their own frogs, as do a fair number of the larger laboratories using genetically engineered *X. tropicalis*.

In general, laboratory females are bred (or eggs collected from them) no more than once per month. The females' expected breeding life ranges from 1 to 5 years, with peak fecundity at 2 to 3 years of age (*X. laevis*). Males can be bred 2 or 3 times per month if necessary. Males have a breeding life expectancy of 3 or more years.

A brief overview on *Xenopus* tadpole rearing is presented here, based on recommendations from several resources (Barnett et al. 2001; Browne and Zippel 2007; Sive et al. 2000) and from http://www.faculty.virginia.edu/xtropicalis/, http://tropicalis.berkeley.edu/home/husbandry/index.html, and http://www.xlaevis.com/tadpoles.html.

- De-jellied, fertilized eggs can be kept in 15–25 gallons of artificial pond water or 0.1 XMMR. During egg development, the tank water must remain chlorine/chloramine free and within a range of 6.5–7.5 pH. The water temperature must approximate the animal's natural habitat and should be in the range of 20° to 23°C (68° to 75°F).

- The water must be adequately aerated and the eggs examined daily with minimal disturbance. A 0.25% malachite green solution (a weak solution) may be used to inhibit fungal growth. Eggs attacked by fungus must be promptly removed (use a disposable, sterile bulb pipette and then discard pipette and eggs after soaking in 10% bleach).

- Approximately 10 to 12 days post-fertilization, hatching larvae can be observed and will normally stay attached to the jelly mass for 1 to 2 days while the yolk sacs are absorbed. Feed the newly developed tadpoles (when tadpoles have reached stage 37/38, ~5 to 10 days after fertilization) with powdered tadpole food (available from frog vendors). Feed tadpoles 2 times per week. Do not overfeed.

- Tadpole stocking density should be approximately 50/L of water initially, and decreased to 5/L when metamorphosis begins. For maximum growth rates, 25–50% of the water should be changed daily after most of the tadpoles have reached the large tadpole and froglet stage (stages 55 and up).

- Tadpoles grow fastest in small groups segregated by size. Larger tadpoles inhibit the growth of smaller ones. Tadpoles should be handled carefully to avoid injury to gills and skin. Tadpoles are relatively fragile and should not be netted. Scooping with containers is recommended. Population density must be continually monitored. Larger froglets should be separated as they will eat the smaller froglets and tadpoles.

- Dead or diseased animals should be collected and necropsied daily.

- Froglets with limb buds (stage 65/66) should receive the same care as adults. Pelleted frog food fed 3 times per week should be gradually introduced and then completely replace the powdered tadpole food. Tank aerators can be removed. Stocking density of 1 frog per gallon will promote rapid growth. *X. laevis* should reach sexual maturity within a year and a half and *X. tropicalis* in 6 to 9 months.

- *Xenopus* are considered "restricted" invasive species in many parts of the world. In the United States, a special permit is required for breeding laboratory *Xenopus* in California and Arizona. Fertilized but unwanted eggs or developing larvae must usually be treated with bleach before they are discarded.

necropsy

The necropsy is the single most important step toward identification of the etiology of disease or adverse effects related to spontaneous conditions, experimental manipulations, water quality, or husbandry management practices. In addition, necropsy will be a critical component of the phenotypic characterization of transgenic frogs.

A careful and complete history is necessary. Pathologists may prefer to euthanize the aquatic species so that autolysis (which is rapid in wet, aquatic specimens held at room temperature) is minimized and the necropsy can be as fresh as possible. If this is not possible, carcasses should be kept at 4°C (refrigeration) until the necropsy is performed. Carcasses should *not* be frozen prior to the necropsy.

Necropsy Equipment

- In addition to personal protective equipment (such as gloves and lab coat), basic equipment required for a *Xenopus* necropsy are small scalpels (scalpel numbers 11 or 15), surgical/dissecting scissors, forceps, small bone cutters, cork or wax board, pushpins, a decalcifying agent such as Cal Ex II®, and 10% buffered formalin in small containers.

- Heart blood, liver, spleen, and coelomic fluid should be aseptically collected for culture as indicated. Therefore, a set of sterile instruments, needles, syringes, culture swabs, appropriate culture transport containers and media, and 70–100% ethanol should always be on hand.

- Additional supplies include tissue bags, tissue cassettes, a standard dissecting scope, good lighting, and a magnifying glass.

Necropsy Technique

- The frog should be weighed, the SVL measured, the gender noted, and the animal examined externally for skin lesions (including wounds, abrasions, and thickening or disruption of the slime coat), petechial and ecchymotic hemorrhages, general body condition, coelomic distension, parasites, fungus, and masses. Skin scrapings and wet mount preps from the skin should be collected if indicated and examined for fungus and parasites.

- The oral cavity should be opened and examined for foreign bodies, abrasions and wounds, and entrapment of a regurgitated stomach.

- The cloaca should be examined for prolapse, engorgement, and hyperemia (as may be seen in sepsis).

- Interdigital webbing should be carefully observed, the digits spread apart, and the webbing examined for congestion of the blood vessels, and petechial and ecchymotic hemorrhages. Interdigital webbing is also one of the first places that small, subcutaneous air bubbles will appear in frogs affected by gas bubble syndrome. These bubbles can best be seen by holding a light behind the foot and illuminating the webbing. A dissecting microscope or magnifying glass may be required.

- The carcass should then be placed in dorsal recumbency on a cork or wax board and pushpins inserted into the feet for stabilization.

- The ventrum of the animal should be disinfected with 70–100% ethanol.

- In *Xenopus*, the skin is loosely attached to the underlying musculature. A large subcutaneous space exists. The skin can easily be reflected back from the midline without incising the abdominal muscles. If excessive subcutaneous fluid is present, collect the fluid for culture and fluid cytology. Bubbles are often present in the subcutaneous space in frogs affected by gas bubble disease.

- The ventral midline musculature is then incised from the pubis to the snout and reflected to expose the coelomic cavity and viscera (**Figure 45**). Coelomic fluid in septicemic frogs may be blood-tinged, opaque, or excessive, and samples should be aseptically collected for culture and cytological evaluation.

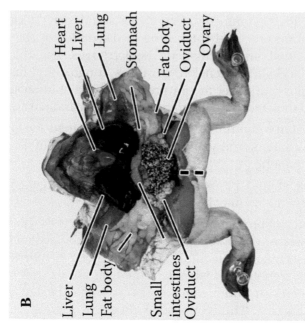

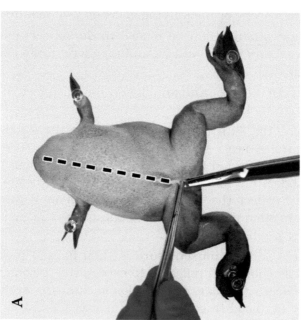

Figure 45 Necropsy of an adult female laboratory *Xenopus*. (A) A ventral midline incision is made to open and expose the coelomic cavity. Major anatomical structures are labeled in B, C, and D.

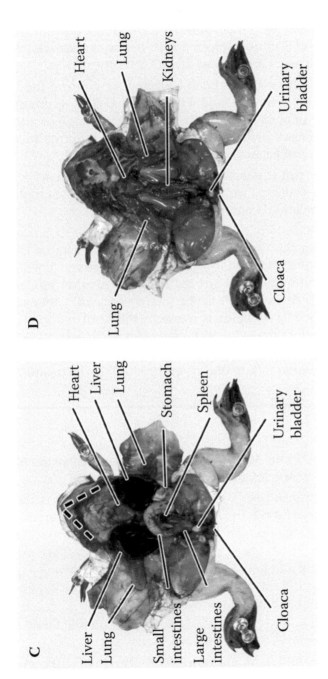

Figure 45 (continued).

- It will be necessary to cut through the sternum and xyphoid cartilage to expose the heart. Heart blood should be aseptically collected for culture, along with aseptically collected samples of liver and spleen. Fresh tissue can also be frozen in liquid nitrogen and stored at –80°C for future use.

- Once specimens have been collected for culture, the viscera should be removed and examined. Collect representative samples of normal and abnormal tissues, and place them in 10% buffered formalin (minimum tissue to formalin volume ratio of 1:10) for histopathology.

- If parasitism is suspected, parasitology can be performed on samples collected from coelomic contents, lung scrapings, or gastrointestinal content and tissue samples.

- It is difficult to remove the brain from small *Xenopus* without damaging the tissue. The bones of the skull can be carefully chipped away, beginning at the foramen magnum, moving along the dorsal midline toward the rostrum. Another method is to remove the whole head at the atlanto-occipital joint and place in Cal Ex II for 48 hours to decalcify the skull and fix the brain in situ before removing it. If using this approach, the eyes will also be fixed while in the skull. The brain can then be trimmed in (with surrounding tissues of the skull) for histopathology.

references

Allender, M. C. and M. M. Fry. 2008. Amphibian hematology. *Vet. Clin. North Am. Exot. Anim. Pract.* 11:463–80, vi.

Animal Research Advisory Committee. 2005. Guidelines for egg and oocyte harvesting in *Xenopus laevis*. http://oacu.od.nih.gov/ARAC/documents/Oocyte_Harvest.pdf (accessed April 28, 2009).

Barnett, S. L., J. F. Cover, Jr., and K. M. Wright. 2001. Amphibian husbandry and housing. In *Amphibian medicine and captive husbandry*, ed. K. Wright and B. R. Whitaker, 35–62. Malabar, FL: Krieger Publishing Company.

Browne, R. K. and K. Zippel. 2007. Reproduction and larval rearing of amphibians. *ILAR J.* 48(3):214–34.

Cannon, M. S. and A. M. Cannon. 1979. The blood leukocytes of *Bufo alvarius*: a light, phase-contrast, and histochemical study. *Can. J. Zool.* 57:314–22.

Chu, D., K. Wright, and P. Sharp. 2006. Hematologic and serum biochemical values for the South African Clawed Frog (*Xenopus laevis*). *J. Am. Assoc. Lab. Anim. Sci.* 45(4):93.

Cross, M. K. and M. Powers. 2008. Obtaining eggs from *Xenopus laevis* females. *J. Vis. Exp.* 20(18):890.

Du Pasuier, L., M. Wilson, and J. Robert. 1996. The immune system of *Xenopus*: with special reference to B cell development and immunoglobulin genes. In *The biology of Xenopus*, ed. R. C. Tinsley and H. R. Kobel, 301–13. Oxford: Clarendon Press.

Duellman, W. E. and L. Trueb. 1986. Morphology. In *Biology of the amphibia*, ed. W. E. Duellman and L. Trueb, 288–415. New York: McGraw-Hill Book Co.

Gentz, E. J. 2007. Medicine and surgery of amphibians. *ILAR J.* 48:255–59.

Green, S. L., R. C. Moorhead, and D. M. Bouley. 2003. Thermal shock in a colony of South African clawed frogs (*Xenopus laevis*). *Vet. Rec.* 152:336–37.

Hadji-Azimi, I., V. Coosemans, and C. Canicatti. 1987. Atlas of adult *Xenopus laevis laevis* hematology. *Dev. Comp. Immunol.* 11:807–74.

Horton, J. D., T. L. Horton, and P. Ritchie. 1996. Immune system of *Xenopus*: T cell biology. In *The biology of Xenopus*, ed. H. R. Kobel and R. C. Tinsley, 279–99. Oxford: Oxford University Press.

Krishnamoorthy, R. V. and N. Shakunthala. 1974. Increased RBC count and pulmonary respiration in cold-adapted frogs. *J. Exp. Biol.* 61:285–90.

Loeber, J., F.C. Pan, and T. Peiler. Generation of transgenic frogs. 2009. *Methods Mol. Biol.* 561:65–72.

Sive, H. L., R. M. Grainger, and R. M. Harland. 2000. *Early development of Xenopus laevis. A laboratory manual.* New York: Cold Spring Harbor.

Sonnen, N. 1970. The effect of endotoxins on frog blood and survival time at several ambient temperatures (34880). *Proc. Soc. Exp. Biol. Med.* 134(3):773–75.

Wright, K. M. 2001. Amphibian hematology. In *Amphibian medicine and captive husbandry*, ed. K. M. Wright and B. R. Whitaker, 129–46. Malabar, FL: Krieger Publishing Company.

resources

---

organizations

1. **The Xenopus Initiative Group:** http://www.nih.gov/science/models/xenopus/. This site posts on progress and plans regarding federal support of the genomic and genetic needs for *Xenopus*-based research. It also provides links to community resources and to information about courses, meetings, and publications.

2. **Xenbase**: http://www.xenbase.org/common/. A *Xenopus laevis* and *tropicalis* genomic and biology community resource.

3. **American Association for Laboratory Animal Science (AALAS),** 9190 Crestwyn Hills Drive, Memphis, TN 38125. Phone (901) 754-8620. AALAS is a professional group serving researchers, animal facility workers, veterinary technicians, and veterinarians. AALAS publishes *The Journal of the American Association for Laboratory Animal Science (JAALAS)* and *Comparative Medicine*. A helpful introductory review of laboratory amphibians can be found on the AALAS Learning Library website at https://www.aalaslearninglibrary.org/demo/course2.asp?strKeyID=C4695509-87A7-4192-AF60-0C6726CD0B78-0&Library=10&Track=11&Series=17&Course=2601&Lesson=27777https://www.aalaslearninglibrary.org/demo/course2.

4. **American College of Laboratory Animal Medicine (ACLAM).** ACLAM is an association of laboratory animal veterinarians founded to encourage education, training, and

research in laboratory animal medicine and is a recognized specialty of veterinary medicine by the AVMA. At the time of this publication, contact is established through the executive director, Dr. Melvin W. Balk, 96 Chester Street, Chester, NH 03036. Phone (603) 887-2467. See http://www.aclam.org/.

5. **Association of Reptilian and Amphibian Veterinarians (ARAV).** 810 East 10th, P.O. Box 1897, Lawrence, KS 66044. Phone (800) 627-0326. The ARAV is a professional association, which publishes the *Journal of Herpetological Medicine and Surgery.* See http://www.arav.org/ECOMARAV/timssnet/common/tnt_frontpage.cfm.

6. **Association for Assessment and Accreditation of Laboratory Animal Care International, Inc. (AAALAC International),** 5283 Corporate Drive, Suite 203, Frederick, MD 21703-2879. Phone (301) 696-9626. AAALAC International is a nonprofit organization that provides a mechanism for peer evaluation of laboratory animal care programs. AALAC accreditation denotes a quality research animal care and use program. See http://www.aaalac.org/.

7. **Institute of Laboratory Animal Resources (ILAR),** 500 Fifth Street NW, Washington, DC 20001. Phone (202) 334-2590. ILAR functions under the National Research Council to develop and make available scientific and technical information on laboratory animals and biologic resources. Through ILAR, such publications as the *Guide of the Care and Use of Laboratory Animals* and the *ILAR Journal* are available. See http://dels.nas.edu/ilar_n/ilarhome/.

8. **Wildlife Disease Association (WDA),** P.O. Box 7065, Lawrence, KS 66044-7065. Phone (800) 627-0326. The WDA is an organization of wildlife biologists, veterinarians, and researchers who study diseases in wild animals and focus on topics such as endangered species, conservation, public health, and comparative medicine. See http://www.wildlifedisease.org/.

electronic resources

Electronic Resources for *X. tropicalis*

For additional information on *X. tropicalis* husbandry, management, research applications, and experimental methodologies:

1. http://tropicalis.berkeley.edu/home
2. http://www.faculty.virginia.edu/xtropicalis/
3. http://www.stjuderesearch.org/depts/pathology/meadlab/husbandry.html
4. http://www.nimr.mrc.ac.uk/devbiol/zimmerman/protocols/
5. http://www.sanger.ac.uk/Projects/X_tropicalis/
6. http://tropmap.biology.uh.edu/
7. http://xtropicalis.cpsc.ucalgary.ca/xmmr/frog.html

Electronic Resources for *X. laevis*

For additional information on *X. laevis* husbandry, management, research applications, and experimental methodologies:

1. http://www.iacuc.arizona.edu/training/xenopus/index.html
2. http://www.petstation.com/clfrog.html
3. http://www.xlaevis.com/
4. http://www.xenopus.com/products.htm, www.xenopus.com, or www.xenopus.com/links.htm
5. http://www-cbd.ups-tlse.fr/organismes/nieuwkoop/nieuwkoop.html
6. http://animaldiversity.ummz.umich.edu/site/accounts/information/Xenopus_laevis.html

Guidelines and Protocols for Harvesting Oocytes

1. http://oacu.od.nih.gov/ARAC/XenopusOocyte_101007_Fnl.pdf
2. http://www.upenn.edu/regulatoryaffairs/Pdf/10HarvestingOocytes.pdf
3. http://tropicalis.yale.edu/obtaining_embryos/squeezing/squeezing.html
4. http://www.xlaevis.com/oocyte.html

Additonal Guidelines

The following agencies provide additional guidelines on husbandry and veterinary care of *Xenopus*.

1. **The Canadian Council on Animal Care (CCAC)** offers species-specific guidelines (CCAC Guide, Volume 2,

1984) and can be found online at http://www.ccac.ca/ en/CCAC_Programs/Guidelines_Policies/GDLINES/ AmphibiansReptiles.htm#amphib.

2. **Amphibians: Guidelines for Breeding, Care and Management of Laboratory Animals, 1996** from the Subcommittee on Amphibian Standards, Committee on Standards, Institute of Laboratory Animal Resources, National Research Council. Go to http://www.nap.edu/catalog.php?record_id=661.

3. **Guidelines for Use of Live Amphibians and Reptiles in Field and Laboratory Research** from the American Society of Ichthyologists and Herpetologists can be found at http://www.asih.org/files/hacc-final.pdf.

4. **Fish, Amphibians, and Reptiles** (National Research Council/ Institute for Laboratory Animal Research). http://dels.nas. edu/ilar_n/ilarjournal/37_4/37_4Amphibians.shtml.

periodicals

The following periodicals publish scientific articles pertaining to laboratory *Xenopus.*

1. *ILAR Journal.* Published by the Institute of Laboratory Animal Resources, National Research Council of the National Academies. See above listing for ILAR.

2. *The Journal of the American Association for Laboratory Animal Science.* Published by the American Association for Laboratory Animal Science. See above listing for AALAS.

3. *Comparative Medicine.* Published by the American Association for Laboratory Animal Science. See above listing for AALAS.

4. *Laboratory Animals.* Published by Royal Society of Medicine Press, 1 Wimpole Street, London W1M 8AE, UK.

5. *Lab Animal.* Published by Nature Publishing Co., 345 Park Avenue South, New York, NY 10010-1707.

6. *Journal of Herpetological Medicine and Surgery.* Published by the Association of Amphibian and Reptile Veterinarians. See above listing for AARV.

7. *Copeia.* Published by the American Society of Ichthyologists and Herpetologists. See above listing for the ASIH.

books

The following textbooks are excellent resources on topics pertaining to *Xenopus.*

1. *A Practical Guide to Creating and Maintaining Water Quality,* 2000 by P. Hiscock, ed. Hauppauge, NY: Barron's Educational.

2. *Early Development of Xenopus laevis: A Laboratory Manual,* 2000 by H. L. Sive, R. M. Grainger, and R. M. Harland. New York: Cold Spring Harbor.

3. *The Biology of Xenopus,* 1996 by R. C. Tinsley and H. R. Kobel. 1996. New York: Oxford University Press.

4. *Color Atlas of Xenopus laevis Histology,* 2003 by A. F. Wiechmann, C. R. Wirsig, and M.A. Norwell, 2003. Kluwer Academic Publishers.

5. *Amphibian Medicine and Captive Husbandry,* 2001 by K. M. Wright, and B. M. Whitaker. Malabar, FL: Krieger Publishing Company.

6. *Environmental Physiology of the Amphibians,* 1992 by M. E. Feder and W. W. Burggren. The University of Chicago Press.

vendor contact information

Microchips

1. Biomark, Inc., 703 South Americana Blvd, Suite 150, Boise, ID 83702. Phone (208) 275-0011, Fax (208) 275-0031. http://www.biomark.com/

2. Microchip ID, 81447 Highway 25, Folsom LA 70437. Phone (800) 434-2843. Fax: (985) 796-1531. http://www.microchipidsystems.com/

3. HomeAgain, registered trademark of Schering-Plough HomeAgainLLC, Kenilworth, NJ 07033. Phone 1-888-466-3242 (1-888-HomeAgain). http://public.homeagain.com/index.html

4. Trovan Electronic ID Devices, Ltd. P. O. Box 40227 Santa Barbara, CA 93140. Phone (805) 565-1288. E-mail: info@eidltd.com. http://www.trovan.com/

5. Visible Implant Elastomer Tags and Visible Implant Alphanumeric Tags, Northwest Marine Technology, P.O. Box 427, Ben Nevis Loop Road, Shaw Island, WA 98286. Phone (360) 468-3375, Fax (360) 468-3844. E-mail: office@nmt.us, www.nmt.us

6. For general instructions on microchip insertion into *Xenopus*, see http://tropicalis.yale.edu/husbandry/tags/E-ANTs.html#ANT.

Carriers Who Will Ship *Xenopus*

1. Menlo Worldwide Logistics, 2855 Campus Drive, Suite 300, San Mateo, CA 94403-2512. Phone (650) 449-1000, (650) 378-5200, Fax (650) 357-9160. https://www.con-way.com/en/logistics

2. DB Schenker (formerly Bax Global), 440 Exchange, P.O. Box 19571, Irvine, CA 92602. Phone (714) 442-4500, (800) 225-5229, Fax (714) 442-2908. http://www.dbschenkerusa.com

3. FedEx. Contact http://fedex.com/us/

4. DHL Express, 1200 South Pine Island Road, Suite 600, Plantation, FL 33324. Phone 800-225-5345.

Frog Suppliers (Frogs and Food)

1. **NASCO INC,** 901 Janesville Avenue, P.O. Box 901, Fort Atkinson, WI 53538-0901. Phone 800-558-9595, Fax 920-563-8296. http://www.enasco.com/page/contact

2. **Xenopus I, Inc.,** 5654 Merkel Road, Dexter, Michigan 48130. Phone (734) 426-2083, Fax (734) 426-7763. http://www.xenopusone.com/

3. **Xenopus Express,** Xenopus Express Inc., P.O. Box 10626, Brooksville, FL 34603-0626. Phone (813) 967-1327; 800-XENOPUS (800-936-6787), Fax (352) 592-7589, info@xenopus.com, http://www.xenopus.com

4. **Aquatic Habitats Incorporated, Inc.,** 2395 Apopka Blvd., Suite 100, Apopka, FL 32703. Orders and General Inquiries: (877) 347-4788, Phone (877) 900-AHAB (2422). Toll-free tech support: (407) 598-1401, Fax: (407) 886-6787, http://www.aquaticeco.com/?gclid=COTB0-PJ5pgCFRxNagoddTNZdA

Modular Housing for Laboratory *Xenopus*

1. **Aquatic Enterprises,** 4101 W Marginal Way SW, Suite A-6, Seattle, WA 98106. Phone (206) 937-0392, Fax (206) 937-1099. E-mail: info@aquaticenterprises.com. http://www.aquaticenterprises.com/index.html

2. **Aquatic Habitats Incorporated, Inc.,** 2395 Apopka Blvd., Suite 100, Apopka, FL 32703. Orders and General Inquiries: (877) 347-4788, Phone (877) 900-AHAB (2422). Toll-free tech support: (407) 598-1401, Fax: (407) 886-6787, http://www.aquaticeco.com/?gclid=COTBO-PJ5pgCFRxNagoddTNZdA

3. **Allentown Inc.,** Corporate Headquarters, P.O. Box 698, Allentown, NJ 08501-0698. Phone (800) 762-2243 or (609) 259-7951. http://www.allentowninc.com/en/

4. **Pharmacal Research Laboratories, Inc.,** P.O. Box 369, Naugatuck, CT 06770. Phone (800) 243-5350 or (203) 755-4908, Fax (203) 755-4309. http://www.pharmacal.com/

5. **Aquaneering, Inc.,** 7960 Stromesa Court, San Diego, CA 92126. Phone (858) 578-2028, Fax (858) 689-9326. E-mail: info@aquaneering.com. http://www.aquaneering.com/

6. **Tecniplast** (headquarters in Italy). Phone +39 0332 809711. E-mail: tecnicom@tecniplast.it, http://www.tecniplast.it/usa/care.php

Sanitation Supplies

1. Virkon® Aquatic can be purchased through Aquatic Habitats Incorporated, Inc., 2395 Apopka Blvd., Suite 100, Apopka, FL 32703. Phone (877) 347-4788, Fax (407) 886-6787. http://www.aquaticeco.com/?gclid=COTBO-PJ5pgCFRxNagoddTNZdA

Water Quality Sensors

1. **YSI, Inc.,** 1700/1725 Brannum Lane, Yellow Springs, OH 45387-1107. Phone (937) 767-7241 or 800-765-4974, Fax (937) 767-9353. E-mail: environmental@ysi.com. https://www.ysi.com/

Water Filtration Systems

1. **Siemens Water Technologies,** 181 Thorn Hill Rd, Warrendale, PA 15086. Phone (866) 926-8420. http://www.water.siemens.com/en/Pages/default.aspx

Water Test Kits (Spectrophotometric)

1. **Hach Incorporated,** P.O. Box 389, Loveland, CO 80539-0389. Phone (800) 227-4224, Fax (970) 669-2932. http://www.hach.com/contact

Water Purifiers

1. **Thermo Fisher Scientific Inc.,** Barnstead Nanopure, 81 Wyman Street, Waltham, MA 02454. Phone (781) 622-1000, Fax (781) 622-1207. http://www.thermofisher.com/global/en/home.asp

taxonomy and natural history

1. Anuran identification (Trent University). A collection of amphibian vocalizations, including keys for identification of both tadpoles and adult frogs. http://www.trentu.ca/biology/berrill/Identification.htm
2. Discussion of the natural history and basic husbandry of the African clawed frog. http://animal.discovery.com/guides/reptiles/frogs/africanclawedfrog.html.

anatomy and histology

1. **DigiMorph** (National Science Foundation). An interactive programming with two- and three-dimensional portrayals of frog anatomy including visual depictions of frog morphology derived from images obtained through computed tomography. http://digimorph.org/specimens/Xenopus_laevis/.
2. **Net Frog** (University of Virginia). A virtual dissection of a frog. http://frog.edschool.virginia.edu
3. **Whole Frog Project** (Lawrence Berkeley National Laboratory). Three-dimensional anatomy of a frog demonstrated through

the use of mechanical sections and MRI images. http://froggy.
lbl.gov

4. **Color Atlas of *Xenopus laevis* Histology**, 2003 by A.
   F. Wiechmann, C. R. Wirsig, and M.A. Norwell. Kluwer
   Academic Publishers.

5. **A color poster of *Xenopus* anatomy** can be purchased
   through AALAS at http://www.aalas.org/bookstore/catego-
   ries.aspx?category=AP.

physiology

1. **Amphibian Biology and Physiology**. A review of amphibian basic biology and physiology. http://digimorph.org/
   specimens/Xenopus_laevis/

2. **Frog morphology and physiology tutorials** (Cornell
   University). A review of amphibian cardiac physiology and a
   photographic atlas of frog anatomy. http://biog-101-104.bio.
   cornell.edu/Biog101_104/tutorials/frog.html.

3. *Environmental Physiology of the Amphibians*, 1992 by M. E.
   Feder and W. W. Burggren: University of Chicago Press.

ontogeny

1. **Developmental staging of *Xenopus*** (Davidson College).
   Photographs that chronicle stages 1–50 in the development of
   Xenopus laevis. http://www.bio.davidson.edu/people/balom/
   StagingTable/xenopushome.html

2. **The normal table of stages of *Xenopus laevis* (Daudin)
   embryonic development.** Reproduced with permission by
   P.D. Nieuwkoop and J. Faber. http://www-cbd.ups-tlse.fr/
   organismes/nieuwkoop/nieuwkoop.html

genetics

1. **Genetic map of *Xenopus*** (University of Houston). A genetic
   map for *Xenopus tropicalis.* http://tropmap.biology.uh.edu

medicine and surgery

1. **Amphibian diseases home page** (James Cook University). Descriptions of diseases documented in wild amphibians and associated with population declines, with emphasis on chytridiomycosis. http://www.jcu.edu.au/school/phtm/PHTM/frogs/ampdis.htm

xenopus listservs

1. http://blumberg-serv.bio.uci.edu/xine/index.htm
2. http://faculty.virginia.edu/xtropicalis/newsgroup.html

index

A

Abrasions and trauma, 109–110
Acariasis, 93–94
Acrylonitrile-butadiene-styrene (ABS)
      pipes, 55, 56
Aeration, 98
Aerobacter spp., 75
Aeromona hydrophila, 75
Aerosolization of disinfectants, 45
Aestivation, 17
Alkalinity, 37–38
Allender, M. C., 131
American Association for Laboratory
      Animal Species (AALAS), 147
American College of Laboratory Animal
      Medicine (ACLAM), 147–148
American Society of Ichthyologists and
      Herpetologists, 150
Ammonia, 33, 39, 41
Amphibian Ringer's solution, 75, 76, 95,
      109, 110
*Amphibians: Guidelines for Breeding,
      Care and Management of
      Laboratory Animals, 1996*, 150
Analgesics, 78, 78–79, 114–115
Anatomic and physiologic features
   aestivation and, 17
   cardiovascular, 14–15
   gastrointestinal and excretory, 10–11
   general, 6
   integument, 7–9
   longevity and, 17
   reproduction, 11–12
   resources, 154–155
   respiratory, 12–14
   sensory, 9–10
   thermoregulation, 16–17
Anemia, 101
Anesthetics, 78, 78–79, 110–112, 117
Animal Care and Use Committees, 63
Animal Welfare Act (AWA), 63
Antibiotics, 77, 108
Antifungal compounds, 77–78
Antiparasitic compounds, 78
Aquaria racks, 20
Aseptic surgery, 112–114
Association for Assessment and
      Accreditation of Laboratory
      Animal Care International
      (AAALAC International), 64, 148
Association of Reptilian and Amphibian
      Veterinarians (ARAV), 148
Azurophils, 133

B

Bacteria, nitrifying, 31–33
Bacterial infections
   antibiotics for, 77, 108
   Chlamydia, 81–83, 84
   Chryseobacterium, 65, 75–80
   Mycobacterium, 65, 80–81, 83
   red leg syndrome, 74–75, 109
Basophils, 133
Batrachochytrium dendrobatidis, 85–87
Behavior, 4–6
Bellerby, C. W., 48
Benzocaine gel, 111–112
Biological filtration, 31–33
Bite wounds, 72, 102–103
Bloating, 72, 73
Blood
   hemocytology, 131–134
   sample collection and interpretation,
      128–131

Bone marrow, 134
Books, 151
Buccal cavity, 13
Buoyancy, 69, 72, 95

C

Canadian Council on Animal Care
    (CCAC), 149–150
Cannibalism, 6
Capillaria xenopodis. See
    Pseudocapillaroides xenopi
Carbon dioxide, 44–45
Cardiocentesis, 129
Cardiovascular system, 14–15
Catching and handling, 125, *126*
Census, 59
Chemical filtration, 33
Chlamydia spp., 81–83, *84*
Chloramines, 23–24, 33, 42–43, 98–100
Chlorine, 23–24, 33, 42–43, 98–100
Chromatophores, 9
Chryseobacterium, 65, 75–80
Citrobacter spp., 75
Clarity, water, 45
Clinical problems, 71–94
    analgesics for, *78–79*, 114–115
    anesthesia and, *78–79*, 110–112
    aseptic surgery for, 112–114
    bacterial infections, 74–83, *84*
    diagnosis, 106–107
    drugs and compounds suggested for
        use in, 75, *77–79*
    fungal infections, 85–89, *90*
    general trauma and abrasions,
        109–110
    noninfectious diseases and
        conditions, 94–106
    parasitic infections, *78*, 90–94
    signs of, 72–74
    surgery for, 112–115
    treatment, 107–110
    viral infections, 84–85
Cloaca, 11, 70, 72
    prolapse, 100–101
Clove oil, 112
Coelomic cavity, 6, 11, 70, 72
Compound administration techniques,
    126–128
Conductivity, 40
Convention on International Trade in
    Endangered Species of Flora and
    Fauna (CITES), 64
Council of Europe, 19
Cryptosporidium, 65, 93

D

Decapitation, 115
Defense against predators, 4
Degassing systems, 98
Dehydration, 94–95
Density, stocking, 47
Desiccation, 94–95
Development and growth, 12, *15–16*
Diet, 5–6. *See also* Food; Nutrition
Dipstick-style quick test color strips,
    46, *48*
Disinfectants, 45
Dissolved oxygen, 43–44, 96–97
Drinking, 9, 10
Dropsy, 72
Drugs and compounds for amphibian
    use, 75, *77–79*
Dry film analyzers, 131

E

E. coli, 75
Ears, 9
Ecchymyosis, 69–70, 72
Eggs, 11–12, *13*, 104–106
    collection and harvesting, 106, 113,
        134–138, 149
Electronic resources, 148–150
Emergency response plans, 46–47,
    49–50
Endangered Species Act (ESA), 64
Enrichment, environmental, 55, *56*
Enterobacter spp., 75
Environmental enrichment, 55, *56*
Eosinophils, 133
Erythrocytes, 132
Eugenol, 112
Euthanasia, *78*, *78–79*, 115–117
Examination, physical, 69–70
Excretory system, 10–11
Experimental methodology
    blood sample collection and
        interpretation, 128–131
    catching and handling *Xenopus* and,
        125, *126*
    compound administration
        techniques, 126–128
    egg/oocyte collection, 106, 113,
        134–138
    hemocytology, 131–134
    interpretation of hemogram and
        serum clinical chemistries, 134
    necropsy, 82, 85, 109, 140–144
    tadpole raising, 138–140

Experimental records, 58
Eyes, 9, 70, 72, 99

F

Federal agencies, 63–64
Fertilization, 11–12
Filtration systems
    biological, 31–33
    chemical, 33
    mechanical, 29, 31
    suppliers, 154
Fish and Wildlife Services (FWS), 64
Flavobacterium, 65, 75
Flow-through systems, 25–27
Food
    amounts, 53–54
    frequency of, 53
    storage, 21
    suppliers, 152
    transportation, 57–58
    types of, 51–53
Freezing, 117
Frequency of feeding, 53
Froglets, 12
Fry, M. M., 131
Fungal infections, 72, *77–78,* 85–89, *90*

G

Gall bladder, 10
Gas bubble disease, 95–98
Gases, total dissolved, 44
Gastrointestinal system, 10–11
Gender differences, 5, 6, 11–12
Genetics, 155
Geography, 2–4
Gout, 70, 101
Granula activate carbon (GAC), 33
Granulocytes, 133
Green, S. L., 115
Growth and development, 12, *15–16*
*Guide for Care and Use of Laboratory
    Animals, The,* 55, 63, 137
*Guidelines for Use of Live Amphibians
    and Reptiles in Field and
    Laboratory Research,* 150

H

Habitat, 2–4
    thermoregulation and, 16–17
Hardness, water, 40–41
Harvesting, egg, 106, 113, 134–138, 149
Health records, 58

Heart, 14–15, 75
    cardiocentesis, 129–130
Hearts, lymphatic, 8
Heavy metals, 45
Hemocytology, 131–134
Hemoglobin, 131
Heterophil blood cells, 133
Hilken, G., 19
Hogben, L., 48
Hormonal priming, 135
Hormones, *79*
Housing
    filtration systems, 29–33
    flow-through systems, 25–27
    macroenvironments, 20–22
    microenvironment, 22–34, *35*
    modular/recirculating systems, 20,
        27–29, *30*
    quarantine, 20, 70–71
    sanitation, 54–55
    static/closed systems, 24–25
    stocking density, 47
    suppliers, 153
    systems and water sources, 22–24
Husbandry
    emergency response plans, 46–47,
        49–50
    environmental enrichment, 55, *56*
    identification, 56–57
    macroenvironment, 20–22
    microenvironment, 22–34, *35*
    nutrition, 51–54
    photoperiod, 21, 48
    record keeping, 58–59
    research and standards, 19–20
    sanitation, 54–55
    stocking density, 47
    transportation, 57–58
    water quality, 35–47
Hydrops, 72, 103–104

I

Identification, 56–57
Illness. *See also* Clinical problems;
        Veterinary care
    diagnosis, 106–107
    drugs and compounds for, 75, *77–79*
    physical examination and, 69–70
    quarantine for, 20, 70–71
    signs of, 72–73
Infections
    bacterial, 71–94, *77,* 108–109
    diagnosis, 106–107
    fungal, 72, *77–78,* 85–89, *90*

general comments on treatment of,
107–109
parasitic, 65, *78*, 90–94
viral, 84–85
Injections, 126–128
Injury and zoonotic risks to personnel,
64–65
Institute of Laboratory Animal
Resources (ILAR), 148
Integument, 7–9
International Air Transportation
Association, 57
Intracoelomic injections, 126, 128
Intramuscular injections, 126, *127*

K

Ketamine, 112
Kidneys, 11, 73, 85, 134
Klebsiella spp., 75

L

Laboratory safety, 64–65
Laparotomy, 137–138
Lateral line system, 10
Lethargy, 75, 82
Light cycles, 21, 48
Listservs, 156
Live Animal Regulations (LAR), 57
Live prey, 52
Liver, 10–11, 75, 83, 134
Longevity, 17
Lucke herpesvirus, 85
Lungs, 12–13
Lungworms, strongyloid, 92–93
Lymphatic hearts, 8
Lymphatic sacs, 7
Lymphocytes, 133–134

M

M. marinum, 65
Macroenvironment, 20–22
Mast cells, 133
McClure, D. E., 115
Mechanical filtration, 29, 31
Medical health records, 58
Melanin pigments, 9, 10
Microchips, 151–152
Microenvironment, 22–34, *35*. *See also*
Water
filtration systems, 29–33
flow-through systems, 25–27
housing systems and water sources,
22–24

modular/recirculating systems, 20,
27–29, *30*
quarantine and, 20, 70–71
static/closed systems, 24–25
UV sterilization, 34, *35*
Migration, 4
Milking, egg, 135–136
Mimi spp., 75
Mites, 93–94
Modular/recirculating systems, 20,
27–29, *30*
Molds, water, 87–89, *90*
Monitoring, water quality, 45–47
Mouth, 10, 70
MS-222, 110–111, 114, 115, 129
Mycobacterium spp., 65, 80–81, *83*

N

National Academy of Sciences, 19, 47, 53
National Research Council (NRC), 63,
64, 150
Natural history, 154
Necropsy, 82, 85, 109, 140–144
Neoplasia, 100
Nitrogen cycle, 31–32, 42
Noninfectious diseases and conditions,
94–106
bite wounds, 72, 102–103
chlorine/chloramine toxicities, 98–100
dehydration and desiccation, 94–95
gas bubble disease, 95–98
gout, 70, 101
neoplasia, 100
ovarian hyperstimulation syndrome,
103–104
poor egg production, poor quality
eggs, 104–106
rectal and cloacal prolapses, 100–101
skeletal deformities, 101
thermal shock, 104
Nostrils, 9–10
Nutrition, 51–54
egg quality and, 105
vitamins and, *79*, 101

O

Occupational health and safety, 64–65
*Occupational Health and Safety in the Care
and Use of Research Animals*, 64
Olfactory lobes, 9–10
Ontogeny resources, 155
Oocyte collection, 134–138
Oral administration of medications, 128
Oral uptake, 9, 10

Organizations, research and
    information, 147–148
Organ meat, 52
Osteoporosis, 101
Ovarian hyperstimulation syndrome,
    103–104
Overcoming the Challenges of Animal
    Transportation, 57
Oxygen
    dissolved, 43–44, 96–97
    saturation levels, 96–97

P

Packed cell volume (PCV), 130
Pancreas, 10
Parasitic infections, 65, *78,* 90–94
Pelleted feed, 51
Percutaneous cardiocentesis, 129
Periodicals, 150
Per os administration of medications, 128
Personnel, laboratory, 64–65
Petechia, 69–70, 72, *74*
pH, 36–37, *39*
Photoperiod, 21, 48
Physical examination, 69–70
Physiology resources, 155
Pithing, 115–117
Potable tap water, 23–24
Predators, 4
Prolapses, rectal and cloacal, 100–101
Propofol, 112
Proteus spp., 75
Pseudocapillaroides xenopi, 70, 90–91, *92*
Pseudomonas spp., 75
Purified water, 24

Q

Quarantine, 70–71
    rooms, 20

R

Ranavirus, 84–85
Record keeping, 58–59
Rectal prolapse, 100–101
Red leg syndrome, 74–75, 109
Regulations and regulatory agencies,
    63–64
Renal system, 11, 73, 85, 134
Reproduction, 11–12, 104–106, 135–136
Resources
    anatomy and histology, 154–155
    book, 151
    electronic, 148–150

genetics, 155
listserv, 156
ontogeny, 155
organization, 147–148
periodical, 150
physiology, 155
taxonomy and natural history, 154
vendor contact information, 151–154
Respiratory system, 12–14
Rhabdias, 92–93
Rickets, 101

S

Sacs, lymphatic, 7
Salmonella, 65
Sanitation, 54–55, 153
Saprolegnia spp., 87–89, *90*
Scoliosis, 101
Sensory features, 9–10
Sexual reproduction, 11–12, 104–106,
    135–136
Shipping containers, 57
Silurana, 1
Skeletal deformities, 101
Skin, 7–9, 11
    dehydrated, 94–95
    gas exchange through, 13–14
    identification by, 56
    illness and, 69, 72, 80, 86, 91, 93–94
    illness diagnosis using, 106–107
Slime layer, 7–9, 72
Spleen, 134
Staphylococcus spp., 75
State agencies, 64
Static/closed systems, 24–25
Sterilization, UV, 34, *35*
Stocking density, 47
Stomach, 10, 70, 107
Stripping, egg, 135–136
Strongyloid lungworms, 92–93
Subcutaneous injections, 126, *127*
Supersaturation, 96
Surgery
    analgesics and post-operative care,
        78–79, 114–115
    aseptic, 112–114
    oocyte harvesting, 136–138
    resources, 156
Swimming problems, 69, 72

T

Tadpoles, 12, *14,* 138–140
Tagging systems, 57
Tap water, potable, 23–24

Taxonomy, 154
Temperature, water, 38–39
  egg quality and, 106
Thermal shock, 104
Thermoregulation, 16–17
Thrombocytes, 133
Torreilles, S. L., 115
Total dissolved gases, 44
Total gas pressure, 96
Transportation, 57–58
Trauma and abrasions, 109–110
Treatment, illness, 107–110
Tricaine methanesulfonate, 110–111,
  114, 115, 129
Trout, 52

U

U. S. Government Principles for the
  Utilization and Care of Vertebrate
  Animals Used in Testing,
  Research and Training (PHS
  Principles), 63
Ultraviolet sterilization, 34, *35*
United States Department of Agriculture
  (USDA), 63
United States Department of the Interior
  (USDI), 64
Urine, 11

V

Vendor contact information, 151–154
Veterinary care
  clinical problems, 71–94
  physical examination, 69–70
Viral infections, 84–85
Virkon Aquatics, 55
Vitamins, *79,* 101

W

Waste materials, 22, 96
Water. *See also* Microenvironment
  alkalinity, 37–38
  ammonia in, *39,* 41

carbon dioxide in, 44–45
chlorine and chloramines in, 23–24,
  33, 42–43
clarity, 45
conductivity, 40
dissolved oxygen in, 43–44
filtration systems, 29, 31–33, 154
hardness, 40–41
miscellaneous toxicants in, 45
molds, 87–89, *90*
nitrogen cycle and, 42
pH, 36–37, *39*
quality, 35–47, 106, 153–154
quality monitoring, 45–47
sources, 22–24
temperature, 38–39
total dissolved gases in, 44
transportation, 57–58
Weight loss, 72, 101
Well water, 23
Wildlife Disease Association (WDA),
  148
Work records, 59
Wounds, bite, 72, 102–103
Wright, K. M., 131
Wright's solution, 131

X

Xenbase, 147
Xenopus epitropicalis, 1
Xenopus Initiative Group, 147
Xenopus laevis, 1–2
  differences between *X. tropicalis*
    and, *3*
  electronic resources, 149
  growth rates, 19
  jumping ability, 25
Xenopus tropicalis, 1–2
  differences between *X. laevis* and, *3*
  electronic resources, 148–149

Z

Zoonotic risks, 64–65